The Blue Revolution

The Blue Revolution
Land Use & Integrated Water Resources Management

WITHDRAWN

Ian R Calder

London • Sterling, VA

First published in the UK in 1999 by
Earthscan Publications Ltd

Moved to digital printing 2004

A catalogue record for this book is available from the British Library

ISBN: 1 85383 634 6

Typesetting by MapSet Ltd, Gateshead, UK
Printed and bound in the UK by Antony Rowe, Eastbourne
Cover design by Yvonne Booth

For a full list of publications please contact:
Earthscan
8–12 Camden High Street
London, NW1 0JH, UK
Tel: +44 (0)20 7387 8558
Fax: +44 (0)20 7387 8998
Email: earthinfo@earthscan.co.uk
http://www.earthscan.co.uk

Earthscan publishes in association with WWF-UK and the International Institute for
Environment and Development.

This book is printed on elemental chlorine free paper

Contents

Figures, Tables and Plates

FIGURES

TABLES

PLATES

Acronyms and Abbreviations

BEA	Bureau of Economic Analysis (USA)
CAQDAS	Computer-Assisted Qualitative Data Analysis Software
CARE	Conservation, Amenity, Recreation and Environment
CGIAR	Consultative Group on International Agriculture Research
CLUWRR	Centre for Land Use and Water Resources Research
CMP	Catchment Management Plan (of NRA, UK)
COM	Council of Ministers
CPSS	Collaborative Planning Support System
CSIRO	Commonwealth Scientific and Industrial Research Organization (Australia)
CUSW	Cebu Uniting for Sustainable Water
DETR	Department of Environment, Transport and the Regions
DFID	Department for International Development (formerly ODA)
DGIS	Netherlands Government Directorate General for International Cooperation
DSS	Decision Support System
DTI	Department of Trade and Industry (UK)
DWAF	Department of Water Affairs and Forestry (South Africa)
EA	Environment Agency (UK)
EPA	Environmental Protection Agency (USA)
FAO	Food and Agriculture Organization (of the United Nations)
GARNET	Global Applied Research Network
GCM	Global Circulation Model
GDP	Gross Domestic Product
GLASOD	Global Assessment of Soil Degradation Project
GNP	Gross National Product
GTZ	Deutsche Gessellschaft für Technische Zusammenarbeit
GUI	Graphical User Interface
GWP	Global Water Partnership
HR	Hydraulics Research
HYLUC97	Hydrological Land Use Change Model, 1997
IAHS	International Association for Hydrological Science
IAWQ	International Association of Water Quality
ICID	International Commission on Irrigation and Drainage
ICOLD	International Commission on Large Dams
ICRAF	International Centre for Research in Agroforestry

ICRISAT	International Crops Research Institute for the Semi Arid Tropics
ICWE	International Conference on Water and the Environment
IEESA	Integrated Economic and Environmental Satellite Account
IFPRI	International Food Policy Research Institute
IH	Institute of Hydrology
IHP	International Hydrology Programme
IIMI	International Irrigation Management Institute
IK	Indigenous Knowledge
IPTRID	International Program on Technology Research in Irrigation and Drainage
IRC	International Water and Sanitation Centre
IUCN	International Union for Conservation of Nature
IWRM	Integrated Water Resource Management
IWSA	International Water Supply Association
LAI	Leaf Area Index
LEAP	Local Environment Agency Plan
LUC	Land Use Change model
NGO	Non-Governmental Organization
NORAD	Norwegian Agency for International Development
NRA	National Rivers Authority
NRBAP	Nile River Basin Action Plan
ODA	Overseas Development Administration (now DFID)
ODI	Overseas Development Institute
PPF	Production Possibility Frontier
PSM	Problem Structuring Method
RSPB	Royal Society for the Protection of Birds
SAC	Structural Adjustment Credit
SAP	Structural Adjustment Programme
SAL	Structural Adjustment Loan
SDR	Sediment Delivery Ratio
SEI	Stockholm Environment Institute
SHE	System Hydrologique Europee
SHETRAN	System Hydrologique Europee Transport
SSSI	Site of Special Scientific Interest
SWIM	System-Wide Initiative on Water Management
TAC	Technical Advisory Committee
TECCONILE	Technical Cooperation Committee for the Promotion of the Development and Environmental Protection of the Nile Basin
TWE	Transferable Water Entitlement
UNCED	United Nations Conference on the Environment and Development
UNDP	United Nations Development Programme
UNEP	United Nations Environment Programme
UNESCO	United Nations Educational, Scientific and Cultural Organization

UNIDO	United Nations Industrial Development Organization
USLE	Universal Soil Loss Equation
VKI	Water Quality Institute (Denmark)
WB	World Bank
WBCSD	World Business Council for Sustainable Development
WEDC	Water Engineering and Development Centre
WHO	World Health Organization
WRC(SA)	Water Research Commission (South Africa)
WRINCLE	Water Resources: INfluence of CLimate change in Europe project
WRSRL	Water Resource Systems Research Laboratory
WSSCC	Water Supply and Sanitation Collaborative Council
WTP	Willingness to Pay
WWC	World Water Council
WWF	World Wide Fund For Nature
ZINWA	Zimbabwe National Water Authority

JOURNALS

Agric for Met	Agricultural and Forest Meteorology
Agric Meteorology	Agricultural Meteorology
For Ecol Manage	Forest Ecology and Management
J App Ecol	Journal of Applied Ecology
J Appl Met	Journal of Applied Meteorology
J Ecol	Journal of Ecology
J Environ Plann	Journal of Environmental Planning and Management
J Geophys Res	Journal of Geophysical Research
J Hydrol	Journal of Hydrology
JIWEM	Journal of the Institution of Water and Environmental Management
Journal of App Ecol	Journal of Applied Ecology
J Royal Soc Western Australia	Journal of the Royal Society of Western Australia
Proc Symp Hydrolog Research Basins	Proceedings of the Symposium on Hydrological Research Basins
Quat J Roy Meteorol Soc	Quarterly Journal of the Royal Meteorological Society
S Afr For J	South African Forestry Journal
Sym Sc Exper Biol	Symposium of Society for Experimental Biology
Water Resour Res	Water Resources Research

Preface

The modern history of water resource and catchment management can be traced from its origins in the achievements of the 19th century engineers whose great civil engineering feats provided wholesome water to the world's growing and industrializing cities. Probably no other single factor contributed as much to the improved quality of life, and life expectancy, of city dwellers as this gift of safe water, and the provision of water-based sanitation systems which then became possible. In those days the surface water catchments, or gathering grounds, were managed to assure the pristine quality of the water. Human occupation was regarded with distrust: it was at best a necessary evil, which had to be contained as far as possible. The success of the engineering approach was not just limited to water supply. The engineer had the ability to 'tame the river'. Through impoundments, barrages and sluices, river flow from catchments could be regulated to reduce floods and to provide more water during times of drought. The Tennessee Valley Authority, which was established in 1933 to control the Tennessee River, a tributary of the Ohio in the USA, perhaps exemplifies best this approach.

More recently the ethos of catchment and water resource management has shifted away from the tightly focused engineering viewpoint. At the same time our perceptions of what we mean by a catchment have subtly changed. Originally 'catchment' may have meant just the headwaters where empoundments had been built to capture water for supply, irrigation and hydroelectricity purposes. Now catchments are regarded more as the hydrological units which occupy the whole land surface of the globe. As demands on water increase this has to be the case, for upland headwater catchments can no longer meet our needs. The need to recycle water, together with the need to exploit groundwater means that more and more we regard every part of the land surface of the globe as part of a catchment which can either supply water or receive our waste water. With this new perception people and the environment can no longer be ignored, and both the human and environmental dimensions are achieving much greater prominence in catchment and water resource management. Indeed the balance has shifted to such an extent that in a recent development project in India (the Karnataka Watershed Development Project), to be funded by the British government's Department for International Development (DFID), although 'watershed' appears in the title virtually no reference to water is made in the project document thereafter. Perhaps this is moving the balance too far?

But this book does not aim to be prescriptive on how the balance should be obtained nor on how water resource management should be carried out. Its purpose is to discuss the issues and to provide new information on land use and water interactions and the tools that are now becoming available, so that those who are involved and affected by water resource management can make the best decisions on how competing demands for water resources in both the short term and long term should be achieved.

The perception of Integrated Water Resource Management (IWRM) that is outlined here is inevitably influenced by the background of the author whose experience lies in research in the physical sciences, management of water resources in a developing country (Malawi), and through acting as the hydrological adviser to a national overseas development organization (ODA, now DFID). The country case studies of the revolution in the way land and water are managed are those which the author knows best; omission of studies does not imply that equally or more important developments may not be taking place in other countries. The perceptions and world-views of economists, ecologists and sociologists, although of course equally valid, are probably not adequately represented here. Clearly for this holistic or systemic approach to land and water management to succeed, these views must be integrated and properly balanced. Other components also need to be included. Health and sanitation, for example, the greatest beneficiaries of the engineering revolution, are already being considered within, and not separate from, integrated water resources management.

Acknowledgements

The author wishes to acknowledge all those who have contributed and are contributing to the 'Blue Revolution'. The ideas presented in this book have arisen from many sources and many disciplines and apologies are given to those whose citations may have inadvertently been omitted. One of the first applications of the term 'blue revolution' known to the author was by Tony Milburn, executive director of the International Association of Water Quality (IAWQ), at the Stockholm Water Symposium in 1996 (Milburn, 1997). At that time it had connotations of increased production; in the context of this book it is intended to indicate the new holistic approach to water management.

The role of South Africa not only in developing an integrated approach to water management as elucidated in the new water bill but in fostering a truly participative approach amongst stakeholders is acknowledged and endorsed.

Particular thanks are given to those who have contributed through discussion and the provision of material to the various case studies that are presented here.

Introduction:
The Revolution

A revolution in the way land and water are managed is under way. There are new appreciations of how land use impacts on water resources and a new recognition that some present land use systems are unsustainable, so that in the medium to long term water resources will be irreversibly degraded. There is a new willingness to reconsider land use and water-resource issues in the light of the ideals and concepts identified at the United Nations Conference on the Environment and Development (UNCED) and in the context of economic Structural Adjustment Programmes (SAP). Together with this new understanding of the land use/water resource system and the new development ideals, new tools and methodologies are being developed to assist the development and execution of water resource management strategies.

Whereas the green revolution provided the means for the developing world to feed itself, the blue revolution addresses much wider issues. These concern both our own generation and future generations as they relate to the long term sustainability of the water resource as it affects not only food production, but also the basic human needs for drinking water and sanitation, for ecology and the environment and for our modern industries and power generation. The green revolution was a technological revolution driven by advances in plant breeding, pest control and the application of fertilizers, its outcome being much greater farm productivity. The blue revolution, although supported by technological advances, is more a philosophical revolution in the way in which we respect the world's environment and one of its most precious assets, water. The outcome of this revolution will be plans and strategies and new designs for land use and all water-related developments.

No attempt is made here to engender another environmental scare story. It is not claimed that without the revolution the world will be doomed within 50 years by entering an ice age or global meltdown, a catastrophic population or resources crisis. The gloomy predictions based on the carrying capacity of world populations in relation to water resources (Falkenmark, 1989) hopefully will not come to fruition. Overstating the dangers may be counter-productive in terms of focusing necessary support. Policymakers and the public are not unaware of the vested interests of environmental scientists and environmental institutions in furthering crisis scenarios. But the need for the revolution is urgent nonetheless. Not in 50 years, but now, there are examples of irreversible environmental degradation, the catastrophic degradation of the Aral Sea and its surroundings being perhaps the most publicized (see UN Department for Policy Coordination and Sustainable Development, 1997). More are in the making, which, without urgent action, will also lead to irreversible decline. Even more are the examples of unsustainable, inequitable and uneconomic land use and water resource systems that need to be addressed.

This book describes the progress of the blue revolution. It outlines some of the new understanding of how land use influences hydrological processes and water resources; it describes the new ideals; it provides examples of case studies where the blue revolution is being applied or needs to be applied and the new methodologies that are being developed to advance it.

The book focuses on the new understanding of the interrelationships between land use and water resources and these are explored in the context of whether much of the widely disseminated folklore and 'mother statements', so often inextricably linked with issues of land use, are based on myth or reality. These myths are so often promulgated by both the media and, perhaps more seriously, by national and international environmental and water-related organizations, that they have permeated and affected land use and water resource planning at the very highest levels.

For two reasons the focus is on the role of forests. This is because forests have very different hydrological properties compared with other land uses. They are often a major land use in upland, mountainous areas which, in many countries, are the wetter sites for replenishing surface water resources. A new approach to estimating the relationship between land use and evaporation and hence on the quantity of water available for runoff or recharge, based on the 'limits' concept, is reviewed.

The new revolutionary ideals, arising from the combination of the UNCED and SAP principles and the new awareness of the need for sustainability, the implications of which are only slowly being appreciated, are outlined.

Examples of the conflicts between land use and water resources in Africa, Asia and Australia illustrate the need and urgency for the new approach.

For conflict resolution, in real world situations where there are many participants with ill-defined objectives, an optimal solution may be neither necessary nor achievable. Using Herbert Simon's (Simon, 1969) concept of 'satisficing', finding a satisfactory solution to all parties, whilst recognizing there may be more than one solution, is perhaps the approach the revolution should adopt.

New methods for linking land-use hydrological models with economics and ecology through decision support systems are outlined and proposed as a framework for the future integrated management of land and water developments at the catchment scale.

Chapter 1

New Understanding: Land Use and Water Interactions

The relationship between land use and water is of interest worldwide. In many developing countries changes in land use are rapidly taking place and the largest change in terms of land area, and arguably also in terms of water resource impacts, arises from afforestation and deforestation activities. Whilst demands for agricultural land and firewood place increasing pressure on the dwindling indigenous forest resource, demands for timber and pulp are leading to increasing areas undergoing commercial afforestation with fast-growing monocultures of often exotic tree species. Agricultural demands for irrigation water compete with those of urban conurbations and industry for water supply. Hydroelectricity, often erroneously thought to have a neutral effect on water resources, because evaporation from reservoirs is not taken into account, is another big user of water in many developing countries, particularly those in southern Africa. For example the flow over the Victoria Falls is approximately equal to the flow from the Kariba dam downstream; all the additional flow into Lake Kariba from the northern-flowing rivers from Zimbabwe and the southern-flowing rivers from Zambia can be considered as lost to evaporation. Expanding industry, increasing urban populations without adequate sewage treatment facilities and greater intensification of agriculture, whilst not significantly affecting the quantity of the resource, all pose problems for its quality.

Land use change not only affects the water resource but may also have long term effects on the land resource. Removal of natural vegetation, overgrazing, poor irrigation and land management methodologies, overexploitation of vegetative cover for domestic use and industrial pollution have all been recognised as causes of soil degradation. Soil degradation may in turn affect the quantity and quality of water as it infiltrates and moves through the soil profile, and alter the hydrological response of catchments and the hydrological cycle.

These are the problems faced by many developing countries when trying to maintain their water and land resources. New approaches to integrated water resources management are being developed. The concepts of demand management and valuing the resource in economic terms that allow competition between higher value uses such as industry and water supply to urban conurbations, as opposed to low value usage such as irrigation, are becoming increasingly accepted.

Knowledge of how land use interacts with the hydrological cycle in relation to evaporation, erosion, land degradation, runoff and infiltration has been built on studies in many parts of the world. These studies use a variety of techniques ranging from geographically large experiments on complete river catchment areas, to detailed examination of the physical and physiological processes operating within the water cycle.

EVAPORATION FROM DIFFERENT VEGETATION TYPES

Although the physics and the physiology of the evaporation process have been the subject of extensive research programmes, and detailed and sophisticated evaporation models have resulted, the outcome of this research exercise has not always been of direct benefit to Integrated Water Resources Management (IWRM). This is because sophisticated evaporation models usually require detailed knowledge of soil and crop parameters and detailed meteorological data which may not be generally available.

Recent advances in our understanding of the differences in evaporation from different vegetation types are reviewed here to provide the basis for the development of an alternative approach to estimating evaporation, outlined in Chapter 3, using the 'limits' concept.

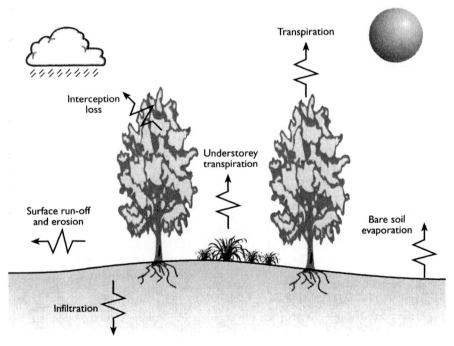

Figure 1.1 *Principal evaporation pathways*

Rainfall over the land surface provides the input for recharging the soil with water, replenishing groundwater reservoirs and providing runoff in streams and rivers. Some of this rainfall will be evaporated into the atmosphere before it can reach either watercourses or groundwater. This evaporation occurs via three pathways: interception, transpiration and evaporation from bare soil. Interception is that proportion of the rainfall held on vegetation surfaces and re-evaporated before it reaches the ground. The water that is drawn up through plant roots and evaporated from the leaves through the stomata (the small pores in the leaf surface) is known as transpiration. The comparative importance of these different pathways is related primarily to the type of vegetation, soil conditions and the climate.

Principal Reasons for Differences in Evaporation between Short and Tall Crops

Forests usually evaporate more water than shorter vegetation or annual crops for two reasons: forests are tall and forest trees generally have deeper root systems. In wet climates, where the surfaces of vegetation remain wet for long periods, interception from forests is higher than that from shorter crops because the very rough surfaces of forests assist the aerodynamic transport of water vapour into the atmosphere. This is analogous to the 'clothes-line' effect, ie, wet clothes pegged out on a line will dry quicker than those laid out flat on the ground. Not only does the increased aerodynamic transport, by reducing the aerodynamic resistance, increase the rate at which evaporated water molecules leave the surface; but it also increases the rate at which heat can be supplied by the atmosphere to the cooler vegetation surface to support the evaporation process. This source of energy, known as advection, is of such significance that annual evaporation rates from forests in wet climates can exceed those that could be sustained by direct radiation from the sun by a factor of two. The difference between the evaporative requirement and the radiant supply is accounted for by the advected energy drawn from the air mass as it moves over the forest.

In drier climates, because forest trees generally have much deeper root systems than short vegetation or agricultural crops, they are able to tap and transpire more soil water during dry periods, and this also leads to higher evaporation rates overall.

Although knowledge of these controls can give a general indication of how the water use of forests may differ from that of other vegetation types, quantifying this difference is still very difficult. This is because there is no simple 'evaporation meter' that can be used to measure the evaporation from the different vegetation types. Also, our knowledge of the complex physical and physiological processes which control evaporation from different vegetation types is still far from complete.

Measurement Methods

There are a number of approaches possible for measuring evaporation from vegetation:

1 Water taken up by roots and transpired through the leaves of plants can be inferred by measuring changes in soil moisture over a period of time.
2 Measurements of the rates of sap flow in the plants themselves can be obtained using tracer methods that use either heat or isotopes of water as the tracer.
3 Measurements of the behaviour of stomata in the leaves, when coupled with information about leaf area and the climatic conditions, can also be used to calculate transpiration rates.

Interception, which is determined by the physical rather than physiological properties of vegetation, is usually measured as the difference in rainfall recorded above and beneath the forest canopy.

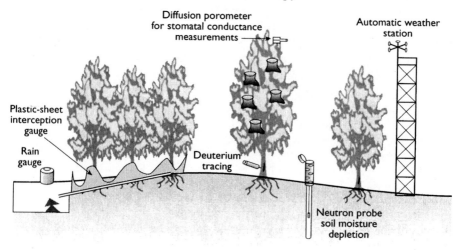

Figure 1.2 *Experimental techniques for measuring evaporation*

LAND DEGRADATION

The term 'land degradation', like the term 'desertification', has no universally accepted definition. The dictionary describes degradation as a loss of strength, efficacy, or value; the wearing away of higher lands. Inevitably the recognition of degradation is subjective. The farmer will recognize degradation where he can no longer grow a crop. Nevertheless it is possible that this 'degraded' land might be providing a more valuable water crop. 'Degradation' of soils and slopes in the mountains, whether naturally occurring or accelerated by man, may be the reason for the fertility of soils in the alluvial plains in the valleys.

Paradoxically 'degraded' and polluted watercourses may often support a wider variety of animal life than when in the pristine state. Deforestation, particularly when it occurs in the tropics, is conventionally and simplistically regarded, almost synonymously, with land degradation.

No attempt is made here to provide a definition of the term 'land degradation' but the warning is given that because any definition will be subjective and necessarily vague, the issue does not lend itself easily to rational and scientific analysis. Often it becomes difficult to discern between myth and reality, and between cause and effect, when dealing with land degradation issues. Conventional wisdom on the processes and mechanisms leading to degradation, and the extent of the problem, needs to be treated with some circumspection. It has also been suggested that obfuscation benefits the vested interests of institutions, consultants and scientists whose existence and livelihoods are dependent on the fostering of crisis scenarios and the design and implementation of what may be economically indefensible amelioration schemes. The wisdom and economic benefits of soil conservation programmes, which have been widely promoted in Africa and Asia, are now under question (Stocking, 1996; Enters, 1998; see also section on water erosion below).

Nevertheless land degradation, in its many guises, is a major issue in land and water resource management and it is urgent that the cause and processes responsible be understood, and the extent of the problem properly assessed, so that it can be addressed within the context of IWRM.

Causes of Land Degradation

As the term land degradation does not lend itself to a simple definition which is generally accepted, it is perhaps not surprising that its fundamental causes also remain obscure and the subject of debate.

Within the developed and industrialized world there can be little doubt that technological advances without adequate controls are a major cause. Here the finger points to industrialization with its consequent pollution and to intensive agricultural practices using either agrochemicals in excess or poorly managed irrigation schemes.

In the developing world the causes are much less clear and perhaps lie more in the realm of socio-economics and sociology rather than technology. Increasing populations and 'population pressure' have been advanced as one of the causes but studies in the Machakos region of Kenya suggest that this is not always the case. Tiffen and co-authors (1994) in their book, '*More People, Less Erosion*', show that in the Machakos increasing population densities in recent years have gone hand in hand with environmental recovery. But these claims have not been universally accepted. Questions have been raised whether the Machakos region is representative of agricultural intensification practices in the rest of Africa and whether it is really on a sustainable growth path.

The conventional view of rural populations being always the cause of deforestation and 'degradation' of savanna areas of Africa, and of greater populations causing increasing deforestation and 'degradation' is also questioned in the book, *Misreading the African Landscape* by Fairhead and Leach (1996). They argue that the forest 'islands' around villages in the savanna region of Guinea are not the remnants of a previous extensive forest cover degraded by an increasing rural population but the result of the rural population actually fostering these 'islands' in a landscape which would otherwise be less woody. The implication here is that population growth has resulted in greater forest cover not less.

The ownership of land, rights of land tenure and poverty have also been put forward as important factors related to degradation. The British government's White Paper on International Development (DFID, 1997) states:

'At the national level, there is a strong link between poverty and environmental degradation. Poor people are often the main direct human casualties of environmental degradation and mismanagement. In rural areas, competition for access to resources, especially land, often squeezes poor people into marginal, low productivity lands, where they have no alternative but to over-exploit soils and forests. In towns and cities, poor people typically have to live and work where pollution is worst and the associated health hazards are highest.'

The White Paper also makes the plea:

'Natural resources must be managed sustainably or else continued economic growth will not be possible. But some use must be accepted or development will not happen. We will help developing countries integrate environmental concerns into their decision-making by supporting their efforts to prepare plans and policies for sound management of their natural resources and national strategies for sustainable development.'

The new understanding that is emerging within the blue revolution is that neither the causes of, nor the solutions to, environmental problems are simple, nor do they rest within one sector or discipline. Solutions can only come within an integrated approach to land and water management.

Extent and Severity of Land Degradation

The Global Assessment of Soil Degradation Project (GLASOD), which was supported financially by The United Nations Environment Programme (UNEP), led to the creation of the world map of the status of human-induced soil degradation (Oldeman et al, 1991). It was a collaborative effort involving 250 soil scientists throughout the world and has been helpful in classifying degradation from a soil scientist's perspective and illustrating the worldwide extent of the problem. The scientists were asked to categorize only soils degraded over the previous 45 years as a result of human intervention. Two categories of human-induced soil degradation processes were recognized. The first dealt with soil degradation through displacement of soil material and the second with chemical and physical soil deterioration. The processes responsible for the first category, soil displacement, were recognized as water and wind erosion. Chemical and physical deterioration were considered to be brought about by loss of nutrients, salinization, acidification and pollution, and by the compaction, waterlogging and subsidence of organic soils (Table 1.1).

Of the total area affected by soil degradation (1964 million hectares), water erosion was responsible for the majority (1094 million hectares). Water erosion, which leads to rill and gully formation, is conventionally associated with the removal of vegetative cover, overgrazing and deforestation activities.

Table 1.1 *Areas and severity of human-induced soil degradation for the world*

Erosion process	Area affected by different degrees of degradation (Mha)					
	Light	Moderate	Strong	Extreme	Total	Total (%)
Water						
Loss of topsoil	301.2	454.5	161.2	3.8	920.3	
Terrain deformation	42.0	72.2	56.0	2.8	173.3	
Total	343.2	526.7	217.2	6.6	1093.7	55.7
Wind						
Loss of topsoil	230.5	213.5	9.4	0.9	454.2	
Terrain deformation	38.1	30.0	14.4	–	82.5	
Overblowing	–	10.1	0.5	1.0	11.6	
Total	268.6	253.6	24.3	1.9	548.3	27.9
Chemical						
Loss of nutrients	52.4	63.1	19.8	–	135.3	
Salinization	34.8	20.4	20.3	0.8	76.3	
Pollution	4.1	17.1	0.5	–	21.8	
Acidification	1.7	2.7	1.3	–	5.7	
Total	93.0	103.3	41.9	0.8	239.1	12.2
Physical						
Compaction	34.8	22.1	11.3	–	68.2	
Waterlogging	6.0	3.7	0.8	–	10.5	
Subsidence of organic soils	3.4	1.0	0.2	–	4.6	
Total	44.2	26.8	12.3	–	83.3	4.2
Total (Mha)	749.0	910.5	295.7	9.3	1964.4	
Total (%)	38.1	46.1	15.1	0.5	4.0	100

Source: Oldeman et al (1991)

Reforestation with plantation forestry is often advocated as the panacea but this may not always be a wise course of action (see also Chapter 2). Wind erosion, also associated with loss of vegetative cover, represents 28 per cent of the affected areas whilst chemical and physical deterioration account for 12 and 4 per cent respectively. The GLASOD programme classified the degree of degradation under four degrees of severity: light, implying some reduction in productivity; moderate, indicating greatly reduced productivity and remedial measures which may be beyond the means of farmers in developing countries; strong, indicating soils that are no longer reclaimable at the farm level; and extreme, which indicated soils that were unreclaimable and beyond restoration.

Erosion by Water

Water is the principal agent of erosion and is responsible for 56 per cent of the world's man-induced soil degradation (Table 1.1). Water erosion occurs mainly as the result of three, sometimes interrelated, processes: sheet erosion, channel erosion and mass movement. In cold climates erosion from water in the frozen form, which occurs when freeze-thaw cycles detach soil particles that are then carried away by rainfall or snowmelt-produced runoff, can also be significant. Erosion is a natural process although it can be accelerated by man's intervention. As described by Newson (1992a)

> '...soils are produced from a bedrock or drift mineral base by weathering and are then eroded as part of the long-term evolution of landscapes. Soil erosion only becomes a problem when its rate is accelerated above that of other landscape development processes – notably weathering – because it becomes visible; it becomes a river management problem when it constrains agricultural production and leads to river and reservoir sedimentation.'

Erosion has different impacts at different scales. At the field or hill-slope scale where 'on-site' erosion may be taking place, material will be removed, while at the larger catchment scale, 'off site' from the source of the erosion, sediments may be being deposited in the river channels, reservoirs and alluvial fans. In many catchments only part of the material that is eroded from the slopes, the on-site erosion, is carried to the stream network. The rest may be held in temporary storage in depressions, foot-slopes, small alluvial fans, behind debris, on flood plains or deposited in the beds of ephemeral channels. It is recognized that as the size of catchments increases the number of storage opportunities for the retention of sediments also increases and it is generally found that the sediment delivery ratio (SDR), the ratio of the on-site erosion to the amount of sediment carried by the stream, decreases markedly with catchment size (Walling, 1983). For small catchments the SDR is typically 0.1, for major catchments, 0.05 and for large river systems, 0.01.

Stocking (1996) shows how, unwittingly or deliberately, those who benefit from soil erosion crisis scenarios have so often used direct scaling, based on area, from the small erosion measurement plot to the catchment scale. In doing so they have conveniently omitted from their calculations the eroded component that is redeposited soon after mobilization, an omission which can lead to overestimates of soil erosion by at least two orders of magnitude. Stocking also highlights other potential dangers associated with small field plot measurements of soil erosion where traditionally soil and water are measured in a trough below a bounded plot with typical dimensions of 20 m upslope by 3 m cross slope. Together with the scaling issue already mentioned, and hence the opportunities for vastly exaggerating the magnitude of the soil erosion problem, he argues that absolute erosion rates from small plot experiments are not, on their own, very helpful for inferring impacts on plant productivity. Erosion-productivity linkages are complex. He cites as examples that on a very susceptible shallow soil, one centimetre of erosion may cause serious reductions in plant yields; on a well drained, high-fertility clay, productivity may be unaffected; yet on a duplex soil, where erosion may expose clays with greater water holding capacity, productivity may actually increase. He also describes the many experimental errors associated with small field-plot measurements, stressing in particular those arising from the intrusive nature of the measurement itself which alters the erosive and deposition processes. Although Stocking concedes that small field-plot erosion studies may have value for determining parameter values for the Universal Soil Loss Equation (see below), for demonstrating the differential effects of land uses and for demonstrating the effects of planting crops in certain ways, the real value of the 'massive erosion research programmes based on small plots' that have been carried out in many developing countries must surely now be in question. Also in question are the economic justifications (Enters, 1998) for many of the major soil conservation programmes that have been advocated for averting many of these soil erosion 'crises'. The quality, and the need for the profusion, of institutions which rely for their existence on propagating conventional amelioration and soil conservation wisdom, should also not escape scrutiny. Stocking (1996) warns that

> *'scientists are just one set of actors in the "soil erosion game", a game in which it is advantageous a) not to admit you do not know the answer; b) to make unverifiable assumptions so that, if your answers provide bad advice, blame does not attach to the professionals; and c) to exaggerate the seriousness of the process to gain kudos, prestige, power, influence and, of course, further work.'*

Sheet Erosion

Sheet erosion begins when drops of water, either raindrops or drops falling from vegetation, strike the ground and detach soil particles by splash. Depending upon the size of the drop and the velocity it has attained, the kinetic energy of the drop, which is released when it strikes the ground, may be sufficient to break the bonds between soil particles and detach them. The movement of water across the ground surface is needed to transport the particles away and complete the process. Without surface runoff the soil erosion losses from nearly level fields are small.

The size of the drop and the distance it falls are crucially important (see below and also Chapter 2). These determine the velocity and the kinetic energy that is achieved. There is also a degree of positive feedback within the process because whenever soil particles are detached, there will be an increase in any surface runoff generation as the finer particles clog soil pores and reduce infiltration.

The erosive potential of rainfall increases with increasing rainfall intensity for two reasons. Firstly, in conditions of high rainfall intensity infiltration rates are more likely to be exceeded and the conditions for generating surface runoff are more likely to be met. Secondly, as rainfall intensities increase, rain drop size also increases. Various equations have been used to describe the relationship between intensity and drop size; one of the first, and one still applicable in most rainfall conditions, is that described by Marshall and Palmer (1948). This empirical equation describes the spectrum of drop sizes that are associated with a particular rainfall intensity, and how the spectra shift with changing intensity (Figure 1.3).

Conventionally, empirical methods such as the Universal Soil Loss Equation or USLE (US Department of Agricultural Research Service, 1961; Wischmeier and Smith, 1965) have been used to estimate sheet erosion rates, at the plot scale, for different land uses. The USLE equation is cast as a simple mutiplicative expression predicting the mass of soil removed from a unit area per annum assuming a knowledge of various causal factors, ie:

$$A = RKLSCP$$

where A is the soil loss per unit area, (usually expressed in tons/acre), R is the rainfall erosivity factor, K is the soil erodibility factor (expressed in tons/acre), L is the field length factor, S is the field slope factor, C is the cropping-management factor normalised to a tilled area with continuous fallow, and P is the conservation practice factor normalised to straight-row farming up and down the slope.

Although the equation was originally devised and calibrated for use in the USA, it has since been used much more widely and various refinements have been introduced to extend the range of calibration (Mitchell and Bubenzer, 1980). However, it is recognized that the equation performs best for medium-textured soils on moderate slopes at the spatial scale of about 100 metres and time scales of a year. When used within its range of calibration, it has proved to be a valuable tool in soil conservation management and can assist in the prediction of annual soil losses under different land uses, and hence aid the selection of cropping and management options and conservation practices.

For erosion prediction over shorter timescales and under variable input conditions, more process-based models of sheet erosion (Morgan et al, 1984) have been developed which separately consider the soil detachment and trans-port processes. The importance of taking into account the role of vegetation canopies as they modify raindrop size and affect rainfall erosivity is discussed below and in Chapter 2.

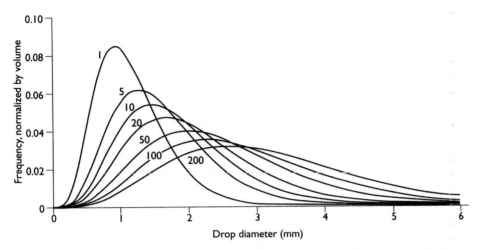

Figure 1.3 *Frequency distribution of rain-drop sizes predicted by the Marshall-Palmer equation, normalized by volume (ie, normalized for equal depths of rain at each intensity) shown for rainfall intensities of 1–200 mm h^{-1}*

Sheet erosion of soils can alter the hydrological functioning of catchments. Where surface infiltration rates are reduced through crusting or the total removal of the soil profile surface, runoff and the likelihood of floods will increase. Recharge to aquifers will also be reduced with concurrent reductions in flows in streams.

This is an area which would benefit from 'new understanding' as the relationship between land degradation and hydrological functioning of catchments remains poorly understood and difficult to assess for particular catchments.

Channel Erosion

Channel erosion, which includes bed and bank erosion, can be a very significant process in natural alluvial channels. It can also be a factor in causing land degradation from channels created by man. These may have been created deliberately as mechanisms for carrying, for example, drainage water from land drainage schemes and roads or accidentally as a result of poor land management practices. Compaction channels formed following land disturbance caused by logging, and crop harvesting in wet conditions, are particular examples. The sediment transport capacity of a channel is generally proportional to the product of the water flow and the channel slope and inversely proportional to the bed or bank sediment size.

Channel formation following land disturbance may also lead to gully erosion. Here waterfall erosion at the gully head, channel erosion in the gully and mass movement of material from the sides of the gully, all erode the gully and drive the gully head upslope.

Mass Movement

Mass movement in the form of landslips is associated with conditions of steep topography and saturated soils and sometimes tectonic movement. Undercutting by rivers may be another factor, but of primary importance are incidences of prolonged and high intensity rains. The role of vegetation in preventing landslips is generally thought to be positive as a result of the binding effect of roots but these benefits are only operative over the rooting depth of the vegetation. Where deep slips occur – usually more serious and taking longer to stabilize – the presence or absence of vegetation does not seem to be a factor. Man's interventions through the building of roads and irrigation canals involving the undercutting of slopes, and in the case of irrigation canals, the saturation of adjacent soils as a result of seepage, have been identified (Bruijnzeel, 1990) as causal mechanisms.

Vegetation, Forests and Erosion

The role of vegetation and particularly forests in relation to erosion is of special interest to IWRM. Conventional wisdom would have us believe that deforestation is often the cause of soil erosion and land degradation and that afforestation is the panacea. Such simplistic views are no longer acceptable. Our understanding of the relationships between forests and erosion has made considerable advances in recent years.

It is now recognized that forests can have both beneficial and adverse effects on erosion. Benefits may result from the binding effects of roots which can prevent landslips on steep slopes and from the generally high rates of infiltration under natural forest which tend to minimize surface runoff. Generally, the adverse effects are associated with poor management practices involving bad logging techniques which compact the soil and increase surface flow, or drainage activities and road construction both of which disturb the soil.

Excessive grazing by farm animals also leads to soil compaction, the removal of understorey plants and greater erosion risk.

Drop Size Modification

Forests can also influence soil erosion by altering the drop size distribution of the incident rainfall. Contrary to popular belief, forest canopies do not necessarily 'protect' the soil from raindrop impacts. In recent years there has been new understanding of the relationship between different tree species, canopies and erosion. Although not generally recognized, the potential for increased erosion from drops falling from forest canopies was demonstrated almost 50 years ago by Chapman (1948). However the importance of species in determining drop size and erosive impacts has not always been well understood. There have been claims that the drop size spectra of drops falling from vegetation are largely independent of vegetation type (Brandt, 1989). The logical consequence of this line of thought is that erosivity would be considered to be unrelated to the type of vegetation.

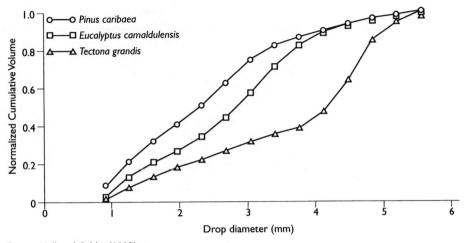

Source: Hall and Calder (1993)

Figure 1.4 *Characteristic net-rainfall drop-size spectra for* Pinus caribaea, Eucalyptus camaldulensis *and* Tectona grandis

However, new theory derived from observations developed from disdrometer observations of the modified drop size spectra beneath canopies of different tree species (Hall and Calder, 1993), suggests a different perspective. These measurements show a well defined, repeatable relationship between the spectrum of drop sizes recorded beneath a particular tree species, termed the characteristic spectrum (Figure 1.4). (Here the spectra are shown as cumulative, rather than frequency spectra, to make it easier to distinguish between spectra.) This same below-canopy spectrum is obtained irrespective of the size spectra of drops incident on the canopy (Figure 1.5).

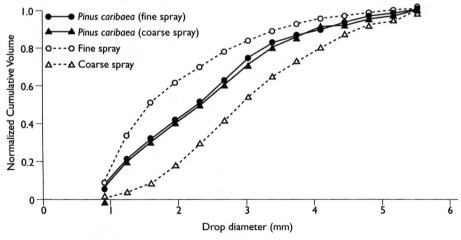

Source: Hall and Calder (1993)

Figure 1.5 *Characteristic net-rainfall drop-size spectra recorded for* P. caribaea, *when subject to fine (1.6 mm median-volume drop diameter) and coarse (2.9 mm median-volume drop diameter) sprays*

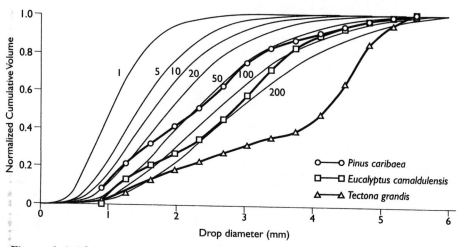

Figure 1.6 *Characteristic net-rainfall drop-size spectra recorded for* P. caribaea, E. camaldulensis *and* T. grandis *and rainfall spectra predicted by the Marshall-Palmer equation for different rainfall intensities (in mm h⁻¹)*

These characteristic spectra can also usefully be compared (see Figure 1.6) with the spectra expected in natural rainfall as predicted by the Marshall-Palmer equation (the rainfall spectra are the same as those shown in Figure 1.3 but redrawn in cumulative form). Clearly the foliage of these tree species will not always reduce the drop size of the incident rainfall. For *Pinus (P.) caribaea*, a rainfall intensity exceeding 50 mmh⁻¹ will be required before any diminution of drop sizes will occur. For *Eucalyptus (E.) camaldulensis* the 'break-even' intensity is about 200 mm h⁻¹ whilst for *Tectona (T.) grandis* it would be 3000 mm h⁻¹, an intensity which could never be reached in natural rainfall.

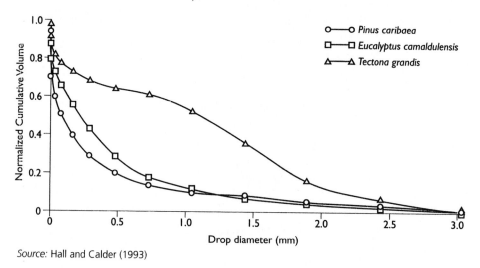

Source: Hall and Calder (1993)

Figure 1.7 *Equivalent drop kinetic energy spectra for* P. caribaea, E. camaldulensis *and* T. grandis *assuming all drops have reached terminal velocity*

Photo IR Calder

Plate 1.1 *Splash-induced erosion under a teak forest in southern India following an understorey fire*

If it is assumed that the height of vegetation is sufficient for all drops to have reached terminal velocity, it is possible to show the characteristic spectra in terms of kinetic energy. In Figure 1.7 the fraction by volume of the sub-canopy drops having kinetic energies exceeding a specified value are shown. This shows how important species differences are in generating drops with different kinetic energies. Median volume drops from *T. grandis* will have nine times greater kinetic energy than those from *P. caribaea*.

In summary the new understanding arising from this work has established that:

1 Below-canopy drop size is independent of the raindrop size falling on the top of the canopy.
2 The below-canopy drop size spectrum is a 'characteristic' of the species.
3 Drop size spectra vary widely between species. This can result in large differences in the potential for erosion; kinetic energies of drops falling from *T. grandis* can be as much as nine times greater than those from *P. caribaea*.

Splash-Induced Erosion: an Observation

Evidence for how severe splash-induced erosion beneath forest canopies can be was provided by observations beneath a teak forest in Karnataka, southern India in 1993. During the dry season of 1993 a fire, a common occurrence in teak plantations, had removed the protective litter layer and understorey

Photo IR Calder

Plate 1.2 *Understorey regrowth protects the soil from drops falling from the leaves of a teak forest*

vegetation. Some regrowth of the understorey had started to take place prior to a severe night-time storm at the outbreak of the monsoon. Major sheet erosion had clearly taken place overnight (Plate 1.1) particularly from the soil which had not been protected by regrowth (Plate 1.2). The 'columns' of soil beneath the protective leaves of the regrowth were approximately 2.5 cm high indicating that, overnight, approximately this depth of soil had been eroded away by sheet erosion.

Observations in other parts of the plantation indicated, from root exposures, that erosion to a depth of two metres had occurred from the plantation since it had been planted about 80 years previously.

These field observations lend dramatic support to the new views on erosion that:

1   For storms with small raindrop sizes (usually those of low intensity) individual drops tend to amalgamate on the surface of a leaf until a large drop is formed which then falls off under the influence of gravity. If the trees are tall, this large drop may reach such a velocity before it reaches the ground that it has both a higher kinetic energy and a higher potential for detaching soil particles than drops in the natural rainfall (see also Chapter 2). In low-intensity storms, therefore, forest canopies will not protect the soil and erosion may be increased.

2   Conversely, for storms with the largest drop sizes, such as in high-intensity convective storms, common in the tropics, vegetation canopies may

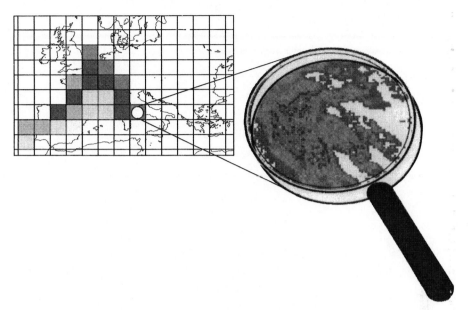

Figure 1.8 *GCMs produce a 'drizzle' over the grid square which needs to be 'downscaled' to produce a more realistic distribution of rainfall in time and space.*

break up the large drops and reduce both the mean drop size and mean kinetic energy of the incident rain. For the highest-intensity storms it would be expected that forests would have an ameliorating effect on erosion, except for large-leafed species such as teak which can never exert an ameliorating effect.

LAND USE, CLIMATE CHANGE AND WATER RESOURCES

The interactions between land use, climate change and water resources present an active research area, but one in which there is little consensus regarding the scale of the effects. Central to the research effort is the use of Global Circulation Models (GCM) to represent the transfer of heat, water vapour and momentum between the surface of the earth and the atmosphere. The GCMs require values for the parameters relating to the surface vegetation type and the availability of water to the vegetation, to allow the estimation of the heat and mass transfer terms, under the atmospheric and rainfall conditions calculated by the GCM. Some of the limitations of the approach, particularly in relation to calculating the impacts of anthropogenic (greenhouse effect) climate change on water resources have been identified by Bonell (1999). Bonell, quoting a recent International Hydrology Programme (IHP) expert group (Shiklomanov, 1999), recognizes that, even given the same initial starting conditions, different GCMs lead to widely differing estimates of the extent of the climate change.

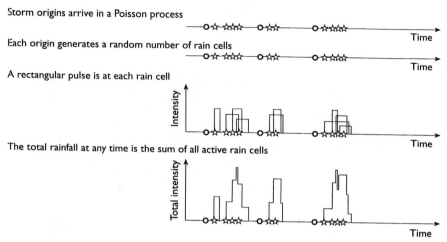

Figure 1.9 *Stochastic rainfall generator uses the generalized Neyman-Scott rectangular pulses model for 'downscaling' in time and space GCM rainfall estimates*

Nevertheless, at present, GCMs represent the best technology available for predicting future climate change and whilst it must be accepted that considerable uncertainty is attached to future GCM predictions there is a need to understand how the range of climate-change scenarios translate to future water-resource scenarios.

Climate change scenarios derived from GCMs are now becoming available for different parts of the world,but there are methodological difficulties in using these to estimate water resource impacts. A central difficulty is that although GCMs are able to simulate present climate in terms of annual or seasonal averages at the continental scale, they are not able to represent the smaller scale features, in time and space, that are more relevant to hydrological and water resource applications. This is because GCMs generate an estimate of the average rainfall over a large (often 50x50 km) grid square for the GCM time step. This 'drizzle' has to be downscaled using knowledge of the spatial/temporal properties of the local rainfall (Figure 1.8). This approach is being adopted for Europe in the WRINCLE project (Water Resources: INfluence of CLimate change in Europe) led by Newcastle University (Kilsby et al, 1998).

Within this project multi-site spatial-temporal rainfall model parameters will be fitted to present climate data using a stochastic rainfall generator based on Poisson-probability statistics (Figure 1.9). The calibrated model will then be used with GCM outputs and an analytical hydrological model to produce discharge statistics. These outputs will then be summarized for Europe in a digital atlas covering precipitation, river discharge and water resource impacts, with respect to the means and absolute and relative changes from the present baseline.

The influence of land use on climate is another area that can be investigated through the use of GCMs and examples of this in Amazonia and the Sahel are given in the next chapter.

Chapter 2

Forests and Water: Myths and Mother Statements

Much folklore and many myths remain about the role of land use and its relation to hydrology, which hinder rational decision making. This is particularly true in relation to forestry, agroforestry and hydrology where claims by enthusiastic agroforesters and foresters are often not supportable. The perception that forests are always necessarily 'good' for the environment and water resources has, however, become so deeply ingrained in our collective psyches that it is usually accepted unthinkingly. The view is routinely reinforced by the media and is all-pervasive; it has become enshrined in some of our most influential environmental policy documents. The report by the UNCED (1992) states:

> 'The impacts of loss and degradation of forests are in the form of soil erosion, loss of biological diversity, damage to wildlife habitats and degradation of watershed areas, deterioration of the quality of life and reduction of the options for development.'

These simplistic views, particularly as they imply the inevitable link between the absence of forests and 'degradation' of water resources, have created a mindset which not only links degradation with less forest but rehabilitation and conservation with more forest. This mindset has caused, and continues to cause, governments, development agencies and UN organizations to commit funds on afforestation or reforestation programmes in the mistaken belief that this is the best way to improve water resources.

Foresters have long been suspected of deliberately propagating some of these forest hydrology myths. Pereira (1989) states in relation to forests and rainfall:

> 'The worldwide evidence that high hills and mountains usually have more rainfall and more natural forests than do the adjacent lowlands has historically led to confusion of cause and effect. Although the physical explanations have been known for more than 50 years, the idea that forests cause or attract rainfall has

persisted. The myth was created more than a century ago by foresters in defence of their trees... The myth was written into the textbooks and became an article of faith for early generations of foresters.'

The overwhelming hydrological evidence supports Pereira's view that forests are not generators of rainfall yet this 'myth', like many others in forest hydrology, may contain a modicum of truth that prevents it from being totally 'laid to rest'.

Swift (1996) has argued, in relation to forests and 'desertification', that the work of E.P. Stebbing, a forester working for the Indian Forest Service in the 1930s, had great influence. Stebbing (1937) promoted the views that the Sahara was both extending and moving southwards, a process more commonly referred to now as 'desertification'. He argued that this extension was the direct result of land use practices, and refers to

'... the present method of agricultural livelihood of the population living in these regions, with their unchecked action of firing the countryside annually, and methods of pasturage – all tend to assist sand penetration, drying up of water supplies and desiccation.'

Reforestation was advocated as the panacea.

When scrutinized, much of the folklore and many of the 'mother statements' relating to forestry and the environment are seen to be either exaggerated or untenable. For others, we still require research to understand the full picture. Seven 'mother statements' in relation to forests, productivity and hydrology are considered:

1   Forests increase rainfall.
2   Forests increase runoff.
3   Forests regulate flows.
4   Forests reduce erosion.
5   Forests reduce floods.
6   Forests 'sterilize' water supplies – improve water quality.
7   Agroforestry systems increase productivity.

Clearly it is important to know what veracity can be attached to these statements for the proper management of water resources and land use. Many forestry projects in developing countries are supported because of assumed environmental/hydrological benefits, whilst in many cases the hydrological benefits may at best be marginal and at worst negative. The evidence for and against each of these 'mother statements' is taken in turn, and appraised; the need for further research is also assessed.

FORESTS INCREASE RAINFALL?

Pereira (1989) denounced the linkage between forests and increased rainfall as a myth yet there may be some situations where this positive linkage cannot be totally discounted and where the presence of forests does lead to a small increase in rainfall. However, as explained later, this small increase in rainfall input will nearly always be more than compensated for by increased evaporation, leading to an overall reduction in water resources. Theory indicates that the height of trees will slightly increase the orographic effect which will, in turn, lead to a slight increase in the rainfall. Modelling studies using mesoscale climate models have shown that some of the intercepted water retained by forest canopies and re-evaporated will return as increased rainfall (Blyth et al, 1994) but this result, although indicating an increase in the gross rainfall above the vegetation, would suggest that the overall net rainfall reaching the ground surface would be reduced as a result of the presence of the forest.

Application of GCMs (Rowntree, 1988) indicates that vegetation changes will have a regional impact on climate. Use of these models in Amazonia shows that total removal of the Amazonian rainforest would affect rainfall patterns with reductions in the rainfall, particularly in the drier north-east of the continent, by about 0.5 mm per day on average (Figure 2.1).

For the whole of the Amazon basin, rainfall would be reduced by six per cent (Institute of Hydrology, 1994).

Similarly, GCM modelling studies for the Sahel (Xue, 1997) indicate that past removal of the indigenous bush vegetation will have altered the spatial distribution of rainfall in a manner that bears a close correlation with observed changes in the distribution patterns (Figure 2.2a and b).

In southern India however, studies of historical rainfall records (Meher-Homji, 1980) indicate that annual rainfall over the last 100 years has *not* decreased despite the large-scale conversion of the dry deciduous forest to agriculture, although there is some evidence of a decrease in the number of raindays.

The issue is well summarized by Bands et al (1987) quoting from experience in South Africa:

> '*Forests are associated with high rainfall, cool slopes or moist areas. There is some evidence that, on a continental scale, forests may form part of a hydrological feedback loop with evaporation contributing to further rainfall. On the Southern African subcontinent, the moisture content of air masses is dominated by marine sources, and afforestation will have negligible influence on rainfall and macroclimates. The distribution of forests is a consequence of climate and soil conditions – not the reverse.*'

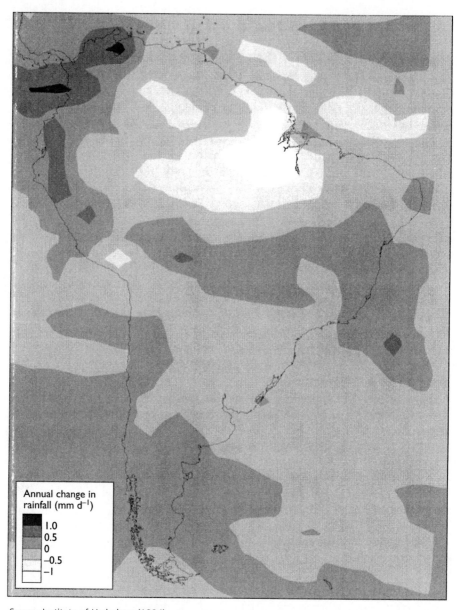

Source: Institute of Hydrology (1994)

Figure 2.1 *Predictions made by the Hadley Centre GCM of the spatial variation of the annual change in rainfall (mm d⁻¹) over Amazonia resulting from complete removal of the Amazon forest*

Conclusion

Although the effects of forests on rainfall are likely to be relatively small, they cannot be totally dismissed from a water resources perspective.

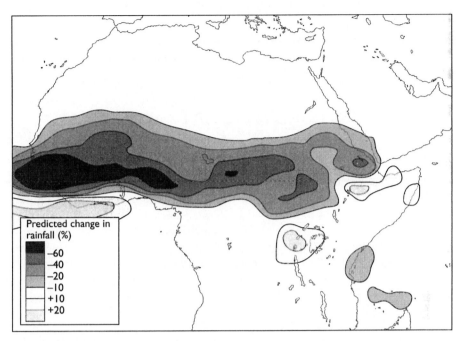

Source: Xue (1997)

Figure 2.2a *The predicted change in rainfall for the months July to September over Central Africa as a result of the degradation of the Sahel vegetation during the last 30 years*

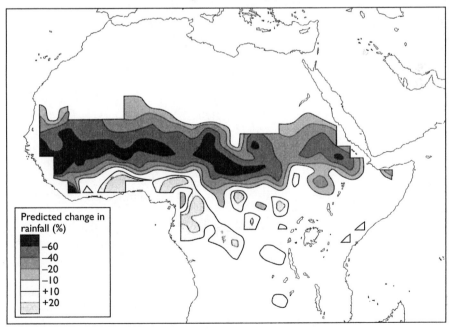

Source: Xue (1997)

Figure 2.2b *Observed change in the rainfall pattern*

Research Requirement

Further research is required to determine the magnitude of the effect, particularly at the regional scale.

FORESTS INCREASE RUNOFF?

A new understanding has been gained in recent years of evaporation from forests in dry and wet conditions based on process studies. These studies, and the vast majority of the world's catchment experiments, indicate decreased runoff from areas under forests as compared with areas under shorter crops. This knowledge has been gained from a host of different studies using a range of different techniques and methodologies. 'Natural' lysimeters have been used to measure total evaporation. Transpiration has been determined using soil moisture measurements (Bell, 1976), micrometeorological and eddy correlation methods (Dyer, 1961), plant physiological studies and tree cutting studies (Roberts, 1977, 1978), and heat, radioactive and stable isotope tracing methods (Cohen et al, 1981; Kline et al, 1970; Luvall and Murphy, 1982; Calder, 1991). Interception has been determined by a number of techniques including interception gauges (Calder and Rosier, 1976), gamma-ray and microwave attenuation methods (Olszyczka, 1979; Calder and Wright, 1986; Bouten et al, 1991), 'wet lysimeters' and rainfall simulators (Calder et al, 1996).

These studies indicate that in wet conditions interception losses will be higher from forests than shorter crops primarily because of increased atmospheric transport of water vapour from their aerodynamically rough surfaces.

In dry (drought) conditions the studies show that transpiration from forests is likely to be greater because of the generally increased rooting depth of trees as compared with shorter crops and their consequent greater access to soil water.

The new understanding indicates that in both very wet and very dry climates, evaporation from forests is likely to be higher than that from shorter crops and consequently runoff will be decreased from forested areas, contrary to the widely accepted folklore.

The few exceptions (which lend some support to the folklore) are:

1   Cloud forests where cloud-water deposition may exceed interception losses.
2   Very old forests. Langford (1976) showed that following a bushfire in a very old (200 years) mountain ash (*Eucalyptus regnans*) forest covering 48 per cent of the Maroondah catchment – one of the water supply catchments for Melbourne in Australia, runoff was reduced by 24 per cent. The reason for this reduction in flow has been attributed to the increased evaporation from the vigorous regrowth forest that had a much higher leaf-area index than the former very old ash forest.
3   Observations and modelling studies of the evaporation from broadleaf forest growing on chalk soils in southern England have been interpreted

as showing reduced water use as compared with grassland (Harding et al, 1992). Further research is planned to investigate these results, which are exceptional in world terms, and to determine if they are applicable to other regions of the UK.

Conclusion

Notwithstanding the exceptions outlined above, catchment experiments generally indicate reduced runoff from forested areas as compared with those under shorter vegetation (Bosch and Hewlett, 1982).

Caveat

Information on the evaporative characteristics of different tree species/soil type combinations are still required if evaporation estimates with an uncertainty of less than 30 per cent are required. In both temperate and tropical climates evaporative differences between species and soil types are expected to vary by about this amount. For example 30 per cent differences in the water use of the same species of *Eucalyptus* growing on different soils have been recorded in southern India (Calder et al, 1993) whilst similar differences have been recorded between different tree species growing on the same soil type also in India (Calder et al, 1997a). In the UK further research is required to determine, at better than 30 per cent uncertainty, the evaporation from different tree species/soil type combinations to establish the potential water quantity impacts of the proposed doubling of UK lowland forests (Calder et al, 1997b).

FORESTS REGULATE FLOWS – INCREASE DRY-SEASON FLOWS?

Although it is possible, with only a few exceptions, to draw general conclusions with respect to the impacts of forests on annual flow, the same cannot be claimed for the impacts of forests on seasonal flow. Different, site-specific, often competing processes may be operating and both the direction and magnitude of the impact may be difficult to predict for a particular site.

From theoretical considerations it would be expected that:

1 Increased transpiration and increased dry-period transpiration will increase soil moisture deficits and reduce dry-season flows.
2 Increased infiltration under (natural) forest will lead to higher soil water recharge and increased dry-season flows.
3 For cloud forests, increased cloud-water deposition may augment dry-season flows.

There are also observations (Robinson et al, 1997) which indicate that for the uplands of the UK, drainage activities associated with plantation forestry increase dry-season flows both through the initial dewatering and in the longer term through alterations to the hydraulics of the drainage system. The importance of mechanical cracking associated with field drainage and its effects on drainage flows has been highlighted by Robinson et al (1985) whilst the work of Reid and Parkinson (1984) indicates that landform and soil type may sometimes be the dominant factors determining soil moisture and drainage flow response.

There are also observations from South Africa that the increased dry-period transpiration is reducing low flows. Bosch (1979) has demonstrated, from catchment studies at Cathedral Peak in Natal, that pine afforestation of former grassland not only reduces annual streamflow by 440 mm, but also reduces the dry-season flow by 15 mm. Van Lill et al (1980), reporting studies at Mokobulaan in the Transvaal, showed that afforestation of grassland with *E. grandis* reduced annual flows by 300–380 mm, with 200–260 mm of the reduction occurring during the wet summer season. More recently Scott and Smith (1997), analysing results from five of the South African catchment studies, concluded that percentage reductions in low (dry season) flow as a result of afforestation were actually greater than the reduction in annual flow. Scott and Lesch (1997) also report that on the Mokobulaan research catchments under *E. grandis*, the streamflow completely dried up nine years after planting; the eucalypts were clearfelled at age 16 years but perennial streamflow did not return for another five years. They attribute this large lag time as being due to very deep soil moisture deficits generated by the eucalypts which require many years of rainfall before field capacity conditions can be established and recharge of the groundwater aquifer and perennial flows can take place.

Bruijnzeel (1990) discusses the impacts of tropical forests on dry-season flows and concludes that the infiltration properties of the forest are critical in how the available water is partitioned between runoff and recharge (leading to increased dry-season flows).

Conclusions

Competing processes may result in either increased or reduced dry-season flows. Effects on dry-season flows are likely to be very site-specific. It cannot be assumed that it is generally true that afforestation will increase dry-season flows.

Caveat

The complexity of the competing processes affecting dry-season flows indicates that detailed, site-specific models will be required to predict impacts. In general the role of vegetation in determining the infiltration properties of soils – as it affects the hydrological functioning of catchments through surface runoff generation, recharge, high and low flows and catchment degradation – remains poorly understood. Modelling approaches, which are able to take into account vegetation and soil physical properties including the conductivity/water content properties of the soil, and possibly the spatial distribution of these properties, will be required to predict these site-specific impacts.

FORESTS REDUCE EROSION?

If foresters are under suspicion for propagating the myth that forests are the cause of high rainfall in upland areas, then there may be equal suspicions raised regarding the oft-cited universal claims of the benefits of forests in relation to reduced erosion. As with impacts on seasonal flows the impacts on erosion are likely to be site-specific, and again, many often-competing processes are likely to be operating.

In relation to beneficial impacts, conventional theory and observations indicate that:

1 The high infiltration rate in natural, mixed forests reduces the incidence of surface runoff and reduces erosion transport.
2 The reduced soil water pressure and the binding effect of tree roots enhance slope stability, which tends to reduce erosion.
3 On steep slopes, forestry or agroforestry may be the preferred option where conventional soil conservation techniques and bunding may be insufficient to retain mass movement of soil.

Adverse effects, often related to forest management activities, may result from:

1 Bad logging techniques which compact the soil and increase surface flow.
2 Pre-planting drainage activities which may initiate gully formation.
3 Wind-throw of trees and the weight of the tree crop reduces slope stability, which tends to increase erosion.
4 Road construction and road traffic which can initiate landslips, gully formation and the mobilization of sediments.
5 Excessive grazing by farm animals which leads to soil compaction, the removal of understorey plants and greater erosion risk.
6 Splash-induced erosion from drops falling from the leaves of forest canopies.

The effects of catchment deforestation on erosion, and the benefits gained by afforesting degraded and eroded catchments will be very dependent on the situation and the management methods employed.

Quoting Bruijnzeel (1990) '*In situations of high natural sediment yield as a result of steep terrain, high rainfall rates and geological factors, little, if any influence will be exerted by man*'. Also in situations where overland flow is negligible, on drier land, little advantage will be gained from afforestation. Versfeld (1981) has shown that at Jonkershoek in the Western Cape of South Africa, land cover has very little effect on the generation of overland flow and soil erosion. On the other hand , in more intermediate conditions of relatively low natural rates of erosion and under more stable geological conditions, man-induced effects may be considerable. In these situations catchment degradation may well be hastened by deforestation and there may also be opportunities for reversing degradation by well-managed afforestation programmes.

Even in these situations, afforestation should not necessarily be seen as a quick panacea. In heavily degraded catchments, such as those on the slopes of the Himalayas, so much eroded material will have already been mobilized that, even if all the man-induced erosion could be stopped immediately, it would be many decades before there was any reduction in the amount of material carried by the rivers (Pearce, 1986; Hamilton, 1987). The choice of tree species will also be important in any programme designed to reduce erosion and catchment degradation.

Recent theoretical developments and observations (see Chapter 1) confirm that drop-size modification by the vegetation canopies of trees can be a major factor leading to enhanced splash-induced erosion. These observations (Figure 1.4) indicate that the degree of modification is species-related, with tree species with larger leaves generally generating the largest drop sizes. The use of large-leaved tree species such as teak (*T. grandis*) in erosion control programmes would therefore be ill advised, especially if there is any possibility of understorey removal taking place.

Conclusions

It would be expected that competing processes might result in either increased or reduced erosion from forests. The effect is likely to be both site- and species-specific. For certain species, for example *T. grandis*, forest plantations may cause severe erosion.

Caveat

Although conventional erosion modelling methods such as the Universal Soil Loss Equation (US Department of Agricultural Research Service, 1961) provide a practical solution to many problems associated with soil loss from agricultural lands, it may not be adequate for the prediction of erosion resulting from afforestation activities. Understanding of the erosive potential of drops falling from different tree species is not adequately appreciated and soil conservation techniques related to vegetation type, soils and slope characteristics have not yet been fully developed.

FORESTS REDUCE FLOODS?

It is a widely held view, propagated by foresters and the media, that forests are of great benefit in reducing floods. Disastrous floods in Bangladesh and northern India are almost always associated with 'deforestation of the

Himalayas'; similarly in Europe floods are often attributed by the media to 'deforestation in the Alps'. However, hydrological studies carried out in many parts of the world – America (Hewlett and Helvey, 1970), South Africa (Hewlett and Bosch, 1984), UK (Kirby et al, 1991; Johnson, 1995) and New Zealand (Taylor and Pearce, 1982) – do not support this view: generally hydrological studies show little linkage between land use and storm flow.

From theoretical considerations it would be expected that interception of rainfall by forests would reduce floods by removing a proportion of the storm rainfall and by allowing the build up of soil moisture deficits. These effects would be expected to be most significant for small storms and least significant for the largest storms.

The high infiltration rates under natural forests also serve to reduce surface runoff and flood response. Certain types of plantation forests may also serve to increase infiltration rates through providing preferential flow pathways down both live and dead root channels. (The use of border trees around agricultural fields subject to surface runoff may provide some prospect for runoff and flood response mitigation whilst not introducing excessive evaporative losses from wide expanses of trees.)

However, field studies generally indicate that it is often the management activities associated with forestry – cultivation, drainage, road construction (Jones and Grant, 1996) and soil compaction during logging – which are more likely to influence flood response than the presence or absence of the forests themselves.

Conclusion

For the largest, most damaging flood events there is little scientific evidence to support anecdotal reports of deforestation as being the cause.

Caveat

Carefully conducted controlled catchment experiments with different climates, soils and species will be required to resolve this issue, but species impacts are probably not as significant as often portrayed. Management activities are most likely to be paramount.

FORESTS 'STERILIZE' WATER SUPPLIES – IMPROVE WATER QUALITY?

Forests were historically the preferred land use for water supply catchments because of their perceived 'sterile' qualities associated with an absence of livestock and human activities. More recently the generally reduced fertilizer and pesticide applications to forests compared with agricultural lands has been regarded as a benefit with regard to the water quality of runoff and recharge. Reduced soil erosion from natural forests can also be regarded as a benefit.

Offsetting these benefits, management activities such as cultivation, drainage, road construction, road use and felling are all likely to increase erosion and nutrient leaching. Furthermore deposition of most atmospheric pollutants to forests is higher because of the reduced aerodynamic resistance of forest canopies compared with those of shorter crops. In high pollution (industrial) climates this is likely to lead to both long-term acidification of the catchment and acidification of runoff.

Conclusions

Except in high pollution climates, water quality is likely to be better from forested catchments. Adverse effects of forests on water quality are more likely to be related to bad management practices rather than the presence of the forests themselves.

Caveat

Studies may still be required to determine the magnitude of the impacts for specific sites and the means to minimize adverse impacts.

AGROFORESTRY SYSTEMS INCREASE PRODUCTIVITY?

Agriculturists have long recognized the productivity benefits that can be gained by mixing different agricultural crops. When a crop such as pigeon pea is mixed with sorghum or maize, much higher production is obtained compared with pure crops of these species which occupy the same total ground area. When productivity of the mixture is superior to that of monoculture it is regarded that the mixture is overyielding and that complementarity has occurred. The Production Possibility Frontier (PPF) is one way in which complementarity and overyielding between crop mixtures can be illustrated (Ranganathan and de Wit, 1996). It can also serve to illustrate the neutral and underyielding situations which can occur between crops due to competition or when one crop inhibits the growth of the other through allelopathic effects (Figure 2.3).

Complementarity can occur when the crops in a mixture, together, can make better use of resources whose supplies are limited. When the resources that crops need for growth (ie, water, light, nutrients and carbon dioxide) are in excess of those needs (ie, in unlimited supply), densely spaced monocultures are generally found to be the most efficient at capturing resources and will have the highest biomass production. Complementarity usually is brought about when one of three conditions is satisfied. Spatial complementarity occurs when the mixture is better able to access resources (such as water, nutrients and light) than the monoculture. Temporal complementarity occurs when the temporal requirements for resources are matched so that high needs in one crop are matched by low needs in the other. The third condition is

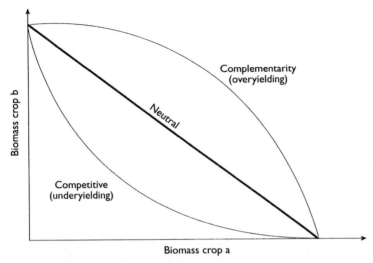

Note: The 'x' and 'y' axes represent biomass production by pure stands of each species and mixtures are shown on the hypotenuse. When the productivity of the mixture lies above the hypotenuse, complementarity occurs; and when below, competition has reduced productivity below that which could be achieved with optimal sole stands of each species.

Source: Ong et al (1996)

Figure 2.3 *Production possibility frontiers (PPF) showing competition and complementarity for two hypothetical crops*

when one of the crops is a legume and can fix nitrogen as a soil nutrient resource which can be of benefit to the other crops.

With the exception of nitrogen fixation, overyielding through complementarity can only come about through greater use of existing limited water or nutrient resources. Most commonly, in arid and semi-arid environments, water will be a resource in limited supply and, for complementarity and overyielding to occur, water use by the mixture will be increased over that of monocultures and less water will be available for other downstream uses whether these be 'environmental' of for supply purposes (see Chapter 4). Within the context of integrated land and water resource management, the benefits of increased productivity need to be assessed in relation to the costs to other possible potential downstream users of the water and 'the environment' as a user, together with the extra labour costs that may be entailed in the sowing, tillage and cropping of a mixture.

It is thought that the increased productivity that has been recorded from mixtures of agricultural crops is largely the result of temporal complementarity, especially when one crop is an annual and the other a biennial, or as a result of nitrogen fixation by one of the crops. Mixtures with pigeon pea, grown either as an annual or biennial legume, exhibit both, and pigeon pea has been found to show complementarity with many species such as maize, sorghum, groundnut and cowpea (Ranganathan and de Wit, 1996).

Agroforestry is built on the belief that mixtures of agricultural crops and trees can also lead to increased productivity, the belief that it is possible to

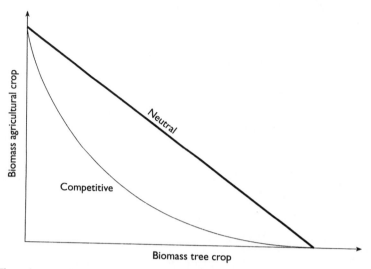

Note: Through competition for resources, particularly water, generally the best that can be achieved is the neutral condition. It is now realized that the presence of the trees will, for all known tree and agricultural crop mixtures, reduce the productivity of the agricultural crop.

Figure 2.4 *Competition and complementarity relationship shown as PPF for a typical agricultural crop and tree crop mixture*

find overyielding mixtures of crops and trees which, when combined, would have a higher yield than when grown separately on the same land area.

A substantial agroforestry literature exists which claims that there are experimental results from agroforestry trials which support this belief. Yet these claims may be erroneous and the trials flawed. From detailed studies of resource capture between both intercropping and agroforestry systems – carried out at the International Crops Research Institute for the Semi Arid Tropics (ICRISAT) sites at Hyderabad in India and the International Centre for Research in Agroforestry (ICRAF) sites at Machakos in Kenya – Ong et al (1991 and 1996) have developed a science framework and process understanding which undermine many of the agroforesters' claims for increased biomass from agroforestry systems. These experimental studies suggest that with agroforestry systems, often the best that can be achieved is near to neutral productivity (Figure 2.4).

By the application of rigorous scientific reasoning, Ong et al (ibid) have also identified some of the pitfalls that have trapped agroforestry researchers into thinking that complementarity and overyielding systems had been achieved. These have arisen mainly because the control plots containing the monocultures have not been under optimal conditions so that the mixture receives a favourable bias.

One identified pitfall is that, for the convenience of statistical analysis, the same plant densities are used in both the monocultures and the mixtures, although, to achieve optimal productivity in the monoculture, higher densities would be required.

A second is that the mixture and monoculture are managed identically even though this management may result in sub-optimal productivity in the monoculture. For example, pruning, which has been used in the mixture to reduce competition from the trees and to return nutrients to the system, would not produce optimal productivity if applied to the control plot.

A third example is that plot sizes may be too small allowing tree roots from the mixture or the monoculture tree plots to penetrate into the plots of the agricultural monoculture and reduce their yields (van Noordwijk et al, 1996). Ong (1996) has shown from studies at Machakos, Kenya, that the roots of the tree *Leucaena leucocephala* can reduce the yield of maize five metres away within two years of growth.

One of the fundamental differences between agroforestry systems and intercropping systems is that the tree component in an agroforestry system, after the initial establishment period, has a well developed and deep root system. Opportunities for spatial complementarity of below ground resources are therefore limited because the tree roots tend to exploit the whole root zone. Furthermore, because, after the initial establishment, the tree roots are always present in the soil profile, there are no opportunities for temporal complementarity. Although trees with nitrogen-fixing root nodules may have potential for complementarity with some agricultural crops, the competitive advantage of trees, which have 'first choice' in tapping soil water and nutrients when the agricultural crops are sown, will usually outweigh these benefits.

Spatial complementarity of above-ground resources, particularly in relation to light capture, is however achievable with tree crop mixtures. Rao et al (1990) have shown that at the ICRISAT site at Hyderabad, a mixture of *Leucaena leucocephala* and millet increased the light capture above that of a sole millet crop. Yet this improved light capture did not result in increased biomass production of the mixture – the biomass produced by the trees was essentially equal to the loss in biomass yield of the crop. This lack of improvement has been explained (Cannell et al, 1998) in terms of the photosynthetic processes operating in trees (all C3 type) which are less efficient in their light to biomass conversion efficiencies than crops, such as millet, which are of C4 type and will have much higher efficiencies. Even though greater resource capture is achieved, this does not translate into higher total productivity.

It is now becoming apparent that trees in agroforestry systems will generally lead to a reduction in biomass of the associated crop (Ong 1996) and neutral total biomass production is usually the best that can be expected.

Modelling studies by Cannell et al (1998) also support this view. Through the use of a process-based agroforestry model which takes into account competition for light and water (but not nutrients), they were able to simulate the growth of a sorghum and tree-crop mixture under different climatic conditions. Their conclusions can be summarized as follows:

1   at sites with less than 800 mm rainfall, maximum total site biomass production was obtained with a monoculture, without overstorey trees;
2   at sites with 800–1000 mm rainfall, neutral biomass production was obtained with a mixture;

3   at sites with greater than 1000 mm rainfall, biomass production would be increased with a mixture provided the Leaf Area Index (LAI) of the trees was greater than 0.25 – but this increase in overall production would be at the expense of a 60 per cent reduction in sorghum grain yields;

4   any decrease in crop yields due to tree competition will automatically increase the frequency of years with poor yields and threaten food security.

Therefore, for low rainfall sites, monocultures would clearly be the best option and, even at higher rainfall sites, to achieve higher total biomass production, it would be necessary to accept a large (60 per cent) reduction in the sorghum crop yield. Cannell et al (ibid) make the point that *'the biomass produced by the trees must be of considerable value, relative to that of sorghum grain, for this sacrifice in yield to be worthwhile'*. They do not give examples of any tree species which would qualify on this account!

Only when soils are very low in nitrogen and the tree crop is a nitrogen fixer, such as *Leucaena leucocephala*, can there be potential for biomass improvement in the associated crop (ICRAF, 1994). Yet when the extra manpower requirements are taken into account in managing agroforestry systems to achieve the complementarity nitrogen fixation advantage, and weighed up against the cost of fertilizer additions to achieve the same end, this benefit may not be seen as being particularly attractive to farmers.

In the more usual situation of soils which are not totally nitrogen deficient, where trees will lead to a reduction in biomass of the associated crop, the reduction in economic productivity is likely to be proportionately much greater. This is especially true for grain producing crops or crops grown in semi-arid, marginal conditions where loss of growth and biomass could lead to total crop failure. Again, when the input costs are taken into account in mixed crop systems, which will be the same or greater than those in monocultures in terms of the cost of seed and the extra manpower requirements, this will further reduce the net production value of the mixed crop as compared with monocultures.

The 'Holy Grail' of agroforestry – a tree species which has roots at a depth which can exploit deep soil resources of water and nutrients but with few roots in the surface layers to offer competition to shallow-rooted agricultural crops – is still being sought, but even if it ever were discovered the spatial complementarity that would be achieved would not be without costs. As for overyielding intercrops, the productivity gains would be at the expense of increased resource use and, when the resource is water, any increased productivity gains would need to be assessed in relation to the (marginal) cost of the extra water consumed.

These recent research findings have far-reaching implications for the practice of agroforestry. They question one of the basic precepts of agroforestry – that optimal mixtures of trees and agricultural crops will lead to an increase in productivity. The results clarify a situation which has been dogged for years by flawed experimental design, obfuscation of results and a singular unawareness that lateral transmission of tree roots can occur over considerable distances and can, when dealing with plot designs of small size, influence control plots. The research shows, contrary to the 'mother statements' underlying much of agroforestry practice, that there are in fact few opportunities for gains in productivity by mixing trees with agricultural crops.

Conclusions

Despite the claims made by over-enthusiastic agroforesters, there is little scientific evidence to show that enhanced productivity can be achieved in agroforestry systems. Growth of the woody component will virtually always be associated with a decrease in biomass and value of the associated annual agricultural crop. Enhanced productivity from agroforestry systems must be largely regarded as a myth. The huge investment in the development and demonstration of agroforestry systems purporting to increase productivity might not have been wasted had more attention been paid by development workers to the indigenous knowledge of local people.

Caveat

Although it would be expected that close proximity competition from the woody component in agroforestry systems would generally prevent productivity benefits, there might well be achievable synergies in agricultural crop and tree crop systems which rely on rotations or tree crop fallows. With these systems it is possible that the deep-rooted nature of most trees may, although consuming deep soil water reserves, bring to the surface through leaf fall, nutrients which are located at depths greater than annual crops can tap. A rotation with an annual crop may then allow these nutrients to be used by the crop whilst the crop itself, through having a less developed root system, will not be able to use all the rainfall, and some water will be available to recharge deeper layers in the soil that the tree crop had previously removed (see reference to sustainable management systems in India, in Chapter 5).

OVERTHROWING THE 'OLD PARADIGM'

The mindset created by the old paradigm which links the absence of forests with 'degradation' of water resources, and 'more forest' with improved water resources, has not yet been destroyed. Until it is replaced it will continue to cause governments, development agencies and UN organizations to commit and waste funds on afforestation or reforestation programmes in the belief that this is the best way to improve water resources.

An example would be in Sri Lanka where in the early-1990s the UK ODA initiated a large-scale forestry programme on the Mahaweli catchments based on the 'old paradigm' assumption that pine reforestation would 'regulate flows and reduce erosion' in the catchments feeding the Victoria reservoir complex. It is now realized (Calder, 1992c) that the pine afforestation is serving only to reduce both annual and seasonal flows. Even if planted at the highest altitudes (where there are still some remaining indigenous cloud forests at Horton Plains) recent research indicates that afforestation would give no net benefit to flows; the measured interception losses from the forest exceed the enhanced cloudwater deposition. Furthermore, the planting is generally taking place on old, degraded tea plantations, where on-site soil erosion has virtually ceased following the regrowth of pattana grasslands. Forestry operations involved with planting and road construction will almost certainly lead to an increase in on-site erosion. Understorey fires under the pines, a common occurrence in Sri Lanka, also leave the soil exposed to splash-induced erosion from the forest canopy. The forestry project is therefore having the opposite effect to that intended. Interestingly Stocking (1996) claims that even where there is bad on-site erosion on the Mahaweli catchment (and erosion from tobacco fields planted on steep slopes is probably the worst source of erosion on the catchment) sediments are not reaching the reservoirs. Stocking claims that the products of the erosion from the slopes are being deposited on the lower slopes and floodplains, and paddy field farmers are actually benefiting from the sediment by incorporating it into their paddies.

Slowly, as the blue revolution takes hold and UN and other development organizations become more aware of the new paradigms relating to land use and water resources (see Chomitz and Kumari, 1998), the pressure will be on national governments to adopt a more questioning attitude to the simplistic 'old paradigm' perceptions about forests and water resources.

Chapter 3

Water Resources and the Limits Concept: a Systems Approach to Estimating Evaporation

The brilliant pioneering work of Howard Penman – originally driven by the needs of the military to know about soil moisture conditions on battlefields which might prevent the movement of tanks and heavy equipment – established a sound physical, climate-related basis for the estimation of evaporation from any surface. Penman's achievement was to solve the three equations which govern the exchange of energy, momentum and heat at an evaporating surface to eliminate the surface temperature term. Through the incorporation of simplifications (the Del approximation – see Calder, 1991) and empirical parameters which related to the properties of the evaporating surface, Penman was able to derive an equation which has been very successful for estimating the evaporation from open water bodies and short vegetation which is not subject to water stress. Monteith (1965), through the incorporation of two terms, the stomatal resistance (the resistance to water movement through stomata) and the aerodynamic resistance (the resistance to water movement from the surface into the atmosphere) provided a framework for estimating evaporation from any surface, whether covered with short or tall vegetation, and whether water limited or not. Evaporation equations, based on the concept of the energy balance and the aerodynamic transport equation (Penman, 1948; Monteith, 1965), are central to most of the modern hydrological methods of estimating evaporation from different surfaces, whether these surfaces are natural vegetation, water or man-made surfaces such as those of urban areas.

These methods have been widely used in models that describe the spatial distribution of evaporation, both in GCMs and in distributed catchment models such as the System Hydrologique Europee (SHE) (Abbott et al, 1986) and more recent developments such as SHETRAN (Ewen et al, 1999).

The obvious success of this approach in many applications has provided little incentive for innovation. Indeed so successful was Penman in promoting his original equation, which implicitly placed heavy reliance on radiation as the primary control on evaporation, that researchers who had observations which were at variance with the equation were either reluctant to publish, or had difficulty in publishing their results. This situation applied to the author.

For many years Penman was unable to accept the results from the Plynlimon forest lysimeter study (see Chapter 5) which indicated that annual evaporation exceeded that given by the Penman equation by 100 per cent.

Neither the physical basis nor the potential accuracy of the Penman-Monteith approach is questioned: it is its practical applicability that is questioned here. For practical applications we are virtually always limited by knowledge of the model parameters, particularly in their spatially distributed form. Furthermore, approaches using the energy balance and aerodynamic transport equations place great emphasis on the importance of these climatic 'demand-led' terms. This may be entirely appropriate in the temperate climates of the world where these evaporation equations were developed, but in many parts of the world, particularly in the very dry regions, the actual evaporation is perhaps only a small fraction of the demand and it may be more reasonable to estimate evaporation from considerations of limits on supply. In other regions other processes may be limiting or more important in controlling evaporation.

The conventional estimation of water use from different vegetation types growing in different parts of the world has required detailed and expensive programmes to measure the evaporation directly or indirectly by measuring evaporation model parameters. An alternative, supplementary approach, however, has been proposed by Calder (1996a and 1998) based on a knowledge of the limiting processes. This approach could be regarded as taking a more holistic or systems perspective on the processes controlling evaporation: less are we concerned with estimating evaporation as some simple or linear function of atmospheric demand; more are we willing to consider the availability of water supply and which scaling (emergent) properties of the vegetation are relevant for estimating evaporation. The manner in which vegetation is wetted is considered, as is vegetation 'size'. It has been suggested (Calder, 1998) that the results from studies carried out in the wet and dry climates of both the temperate and tropical world can be interpreted in terms of six types of controls and limits on the evaporation process: advection, radiation, physiology, soil moisture, tree size and drop size. From knowledge of the climatic zone and the type of crop, it is claimed that 'broad-brush' estimates of water use, of sufficient accuracy for many IWRM requirements, can be obtained. Water use measurements in the different climatic zones and from different vegetation types are reviewed in the context of the limits concept below.

WET TEMPERATE CLIMATE: SHORT AND TALL CROPS

The results from studies carried out at the Plynlimon experimental catchments in mid Wales, the Balquhidder experimental catchments in central Scotland and at the Crinan Canal catchments in the west of Scotland, illustrate some of the important controls on evaporation from vegetation growing in these wet temperate climates (Calder, 1986, 1990 and 1992a; Kirby et al, 1991; Johnson, 1995).

Plate 3.1 *A neutron probe being used on the Plynlimon forest lysimeter*

Advection Limit

The Plynlimon 'natural' plot lysimeter (Plate 3.1) contains 26 spruce trees, hydraulically isolated by containing walls and impermeable clay subsoil, and clearly demonstrates the importance of the interception process for upland forest.

Annual forest interception losses, determined by large plastic-sheet net-rainfall gauges, were about twice those arising from transpiration, determined from the lysimeter and neutron probe measurements. The total evaporative loss, from both transpiration and interception required a latent heat supply, which was supplied by large-scale advection, and which exceeded the radiant energy input to the forest (Table 3.1, page 48).

The uplands of the UK, subject to a maritime climate typified by high rainfall, a high number of raindays per year and high windspeeds, are an example of a situation where large-scale advection of energy routinely occurs from moving air masses as they pass over wet forest covers. The high aerodynamic roughness of forest compared with shorter vegetation types allows the transport of heat to the forest surface from the air, and the transport of water vapour from the surface into the atmosphere, to occur at rates up to ten times higher than those possible from shorter vegetation. The use of advected energy is therefore much higher for forest and is the principal reason for the much higher evaporative losses from upland forests compared with shorter vegetation types. In the UK uplands advection can probably be regarded as

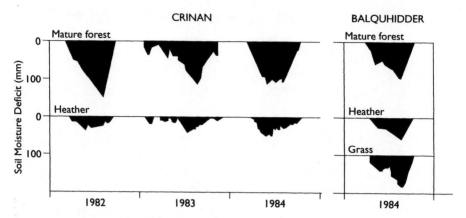

Note: Measurements made with access tube networks located to the full rooting depth beneath mature spruce forest, heather and grass moorland

Source: Calder (1990)

Figure 3.1 *Soil moisture deficits determined from neutron probe measurements at Crinan and Balquhidder*

not only a major source of energy for forest evaporation but as the principal limit on the evaporative process.

Radiation and Physiological Limits

Shorter vegetation is less able to draw on advective energy to augment evaporation rates. For shorter vegetation, aerodynamic roughness is less, and evaporation rates are more closely linked to the supply of radiant energy, ie, they are radiation-limited. Stomatal controls, ie, physiological controls on transpiration, also become more important. Soil moisture deficits recorded under heather, *Calluna vulgaris*, grass and coniferous forest, at the Balquhidder and Crinan sites are shown in Figure 3.1.

Modelling studies (Calder, 1990) indicate that at these wet upland sites with an annual rainfall in excess of 1500 mm and with vegetation growing on generally deep peaty soils, soil moisture availability is not usually a limit on evaporation. The differences in the soil moisture deficits under heather and grass are principally a reflection of the increased physiological controls on transpiration imposed by heather as compared with grass. The much higher deficits recorded under forest are again a demonstration of the overriding importance of interception in determining forest evaporation in these climates.

DRY TEMPERATE CLIMATE: SHORT AND TALL CROPS

Physiological and Soil Moisture Limits

Measurements of the evaporative and soil moisture regime under ash and beech forest in southern England were included as part of an investigation into the hydrological impact of increased areas of broadleaf plantations (Harding et al, 1992).

At these sites evaporation was strongly influenced by physiological controls on transpiration imposed by the trees and soil moisture availability and, to a lesser extent, interception. Differences in soil moisture availability were strongly related to soil type. For ash and beech growing on soil overlying chalk the available soil water was essentially infinite, whereas at the clay-soil site the available water to the trees was of the order of 280 mm (Figure 3.2)

From measurements of stomatal conductance, Harding and colleagues concluded that the physiological controls on the transpiration from the beech forest were sufficient to reduce the total annual water use to less than that expected for grass. However, this conclusion, that the water use of the forest is less than that from a short crop, has recently been questioned and is the subject of current investigations that are described in Chapter 5.

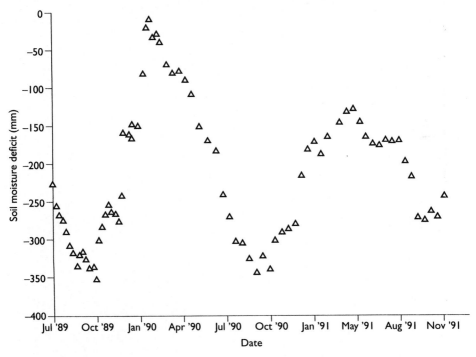

Source: Harding et al (1992)

Figure 3.2 *Observations of soil moisture deficits exceeding 350 mm under beech on a chalk site at Black Wood in southern England*

Radiation and Soil Moisture Controls

For grassland and other shorter crops it is generally recognized that radiation controls, together with soil moisture controls, are the major determinants of evaporation. The Penman approach (Penman, 1963) – which takes account of the radiation control within the calculation of potential evaporation, and the soil moisture control within a 'root constant' function – has been shown to be quite adequate for calculating evaporation from short crops in generally dry temperate conditions.

Dry Tropical Climate: Short and Tall Crops

Soil Moisture

As part of a study to investigate the hydrological impact of eucalyptus plantations in the dry zone of Karnataka, in southern India, comparative studies were carried out on the evaporative characteristics and soil moisture deficits under eucalyptus plantation, indigenous forest and agricultural crops. For all of the four sites where investigations were carried out, soil moisture availability was found to be a major limit on evaporation for both agricultural crops and trees. For finger millet (*Eleusine coracana*), the annual agricultural crop studied, the rooting depth was less than two metres and the available water was found to be 160 mm. This compares with 390 mm for most of the forest sites.

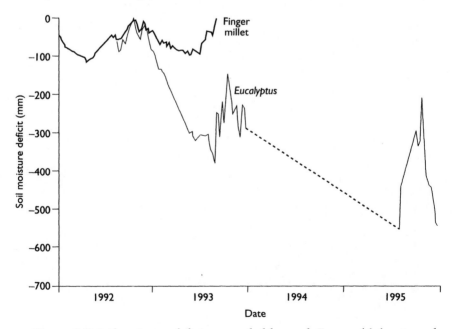

Figure 3.3 *Soil moisture deficits recorded beneath* E. camaldulensis *and finger millet at the Hosakote site, India*

At one of the eucalyptus sites, the Hosakote experimental site of the Karnataka Forest Department, the roots are now known to extend to much greater depths. Recent studies using neutron probe measurements made to a depth of eight metres on a 'farmer's-field' experiment (Calder et al, 1997a), have shown huge differences in the soil moisture deficits recorded beneath finger millet and eucalyptus (Figure 3.3)

The results (Figure 3.4) show that not only can eucalyptus roots reach the lowest measurement depth of eight metres but that they can reach this depth within two to three years of being planted. Together with evaporating essentially all the rainfall that enters the soil, these trees are able to extract approximately an additional 100 mm of water from each metre depth of soil the roots penetrate. The concept of soil moisture availability cannot easily be applied to the eucalyptus plantations at this site.

The deep-rooting behaviour of *Eucalyptus* species has also been reported in South Africa, where Dye (1996) found *E. grandis* abstracting from a depth of eight metres.

Nevertheless for the agricultural crops it is clear that reduced soil moisture availability was the principal reason why the annual evaporation from these crops was generally about half that from either plantation or indigenous forest.

Tree Size

These Indian studies also demonstrated, for the relatively young (less than seven years old) plantation trees used in the study, a linear scaling relationship between transpiration rate and basal cross sectional area (Figure 3.5). For

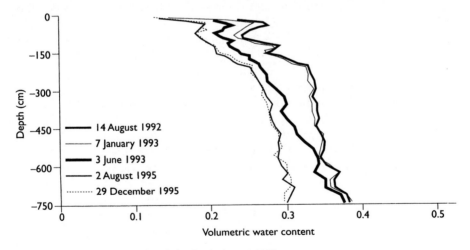

Note: Observations from the day of planting in August 1992.

Source: Calder et al (1997a)

Figure 3.4 *Neutron probe observations of profile volumetric water contents beneath* E. camaldulensis *at Hosakote, India*

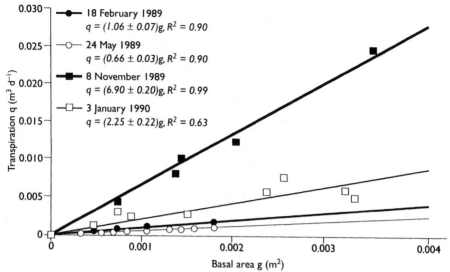

Note: Prior to and after the monsoon in July–September 1989.

Source: Calder (1992b)

Figure 3.5 *Measured transpiration rates as a function of tree basal area at Puradal, India*

older plantations (greater than ten years old) where, through competition, self-thinning was taking place, the relationship was less well defined (Calder et al, 1992).

It would appear that although evaporative demand is clearly the driving mechanism for evaporation, for most of the year it is not limiting; the primary controls are soil moisture availability, and for the tree crops, some factor relating to tree size. At the sites in India, which experience an extended dry season, interception losses, which amount to less than 13 per cent of the annual 800 mm rainfall, are not important in determining soil moisture deficits and are not a major component of the total evaporation.

The results from semi-arid Karnataka (Calder et al, 1993), which indicate that evaporation is limited principally by soil water availability and plant physiological controls, are therefore in direct contrast to the observations from the wet uplands of the UK where evaporation is principally limited by atmospheric demand through advection and radiation controls.

WET TROPICAL CLIMATE: TALL CROPS

Drop Size Controls

At wetter sites in the tropics, interception is a more significant component of the annual forest evaporation. From studies carried out in rainforest in Indonesia (Calder et al, 1986) the importance of raindrop size in determining

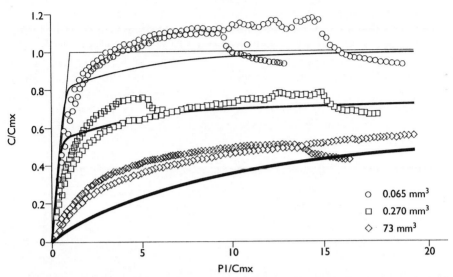

Note: The storage of water on the sample, C, is shown normalized by the maximum storage obtained with fine drops, Cmx, and the depth of simulated rainfall applied, P1, is also normalized by Cmx. Also shown as continuous lines, with the line thickness increasing with drop size, are the wetting functions predicted by the stochastic interception model with model parameters: v_e = 4 mm^3, v_c = 11.5 mm^3, L^*_p = 0.799. The wetting function implicit in conventional interception models of Rutter type are shown as straight lines.

Source: Calder et al (1996)

Figure 3.6 *Canopy storage measurements obtained with a rainfall simulator showing that the final degree of canopy saturation is much greater when wetted with drops of smaller size*

interception losses from tropical forest was first realized. Application of a stochastic interception model, which explicitly took into account drop size, was required to describe the interception process in these conditions. This model shows that up to ten times as much rain may be required to achieve the same degree of canopy wetting for tropical convective storms, with large drop sizes, than would be necessary for the range of smaller drop sizes usually encountered for frontal rain in the UK. There are also results from studies using rainfall simulators, which show that the final degree of canopy saturation also varies with drop size, being greater for drops of smaller size (Figure 3.6).

Vegetation canopies also influence the drop size of the net rainfall: deeper layers in the canopy will be more influenced by the modified drop size spectrum falling from canopy layers above, than by that of the incident rain. Recent studies (Hall and Calder,1993) have shown that different vegetation canopies have very distinct canopy spectra (Figure 1.5). For canopies with a low leaf-area index, the interception characteristics would therefore be expected to be related more to the drop size of the incident rain, whereas for canopies with a higher leaf-area index, the characteristics would be less dependent on the drop size of the rainfall.

The advantages to be gained by incorporating the drop size dependence in interception models for use in tropical conditions were demonstrated by Hall et al (1996a) for a tropical forest site in Sri Lanka. The performance of the two-layer stochastic interception model (Calder, 1996b; Calder et al, 1996), which explicitly takes into account drop size, was very much better in describing the initial wetting-up phase of the storm, and hence the overall interception loss, than conventional interception models (Rutter et al, 1971). The drop size dependence of canopy wetting provides part of the reason why forest interception varies so much worldwide, and why interception losses from coniferous temperate forests are so much higher than from tropical forests (Table 3.1).

Table 3.1 *Observations of the annual water use and energy balance of wet temperate and wet tropical forests*

Site	Rain (mm)	Transpiration (mm)	Interception (mm)	Total Evaporation (mm)	Net Radiation (mm equiv.)
Wet Temperate Plynlimon, Wales 1975	2013	335	529 (26%)	864	617
Wet Tropical West Java Aug 80–July81	2835	886	595 (21%)§	1481 ±12%	1543 ±10%
Wet Tropical Ducke Forest, Brazil	2593	1030	363 (14%)	1393	1514

Note: The Ducke Forest site experiences dry periods which may limit transpiration.
Source: Calder (1978), Calder et al (1986), Shuttleworth (1988)

The model shows that canopy wetting will be achieved most rapidly and maximum canopy storage will be highest, leading to high interception losses overall, when the volume of individual raindrops and drops draining from the canopy, are both small. These conditions apply for coniferous forests in the low intensity, small raindrop size climate of the uplands of the UK. By contrast, when both individual raindrop volumes and leaf drop volumes are large, canopy wetting will be achieved much more slowly, the final degree of canopy saturation will be less, and interception losses are likely to be much reduced. This situation is typified by tropical rainforest experiencing high intensity convective storms of large drop size.

Radiation Limit

The wet evergreen forests of the tropics represent another situation where climatic demand is likely to limit forest evaporation. However, climate circulation patterns in the tropics do not favour large scale advection of energy to support evaporation rates, and here evaporation rates are likely to be closely constrained by the availability of solar radiation. Whereas in the wet temper-

Table 3.2 *Annual evaporation and precipitation for selected tropical lowland forests*

Site and Source of Information	Precipitation (mm.yr^{-1})	Evaporation (mm.yr^{-1})	Length of observations (yr)
Latin America			
Ducke Reserve, Brazil (Shuttleworth, 1988)	2468	1311	2
Ducke Reserve, Brazil (Leopoldo et al, 1982b)	2075	1675	1
Bacio Modelo, Brazil (Leopoldo et al, 1982a)	2089	1548	1
Tonka, Surinam (Poels, 1987)	2143	1630	4
Gregoire I (Roche, 1982)	3676	1528	8
Gregoire II (Roche, 1982)	3697	1437	8
Gregoire III (Roche, 1982)	3751	1444	8
Barro Colorado, Panama (Dietrich et al, 1982)	2425	1440	2
Idem	1684	886	1
Africa			
Tai I, Ivory Coast (Collinet et al, 1984)	2003	1465	1
Tai II, Ivory Coast (Collinet et al, 1984)	1986	1363	1
Banco I, Ivory Coast (Huttel, 1975)	1800	1145	3
Banco II Ivory Coast (Huttel, 1975)	1800	1195	2
Yapo, Ivory Coast (Huttel, 1975)	1950	1425	1
Guma, Sierra Leone (Ledger, 1975)	5795	1146	3
Yangambi, Congo (Focan and Fripiat, 1953)	1860	1433	1
South-East Asia			
Sungai Tekam C, Malaya (DID, 1986)	1727	1498	6
Sungai Tekam B, Malaya (DID, 1986)	1781	1606	3
Sungai Lui, Malaya (Low and Goh, 1972; DID, 1972)	2410	1516	3
Angat, Phillipines (Baconguis, 1980)	3236	1232	6
Babinda, Queensland (Gilmour, 1977; Bonell and Gilmour, 1978)	4037	1421	6
Janlappa, Indonesia (Calder et al, 1986)	2851	1481	1

Source: Bruijnzeel (1990)

ate, maritime climate of Plynlimon in mid Wales, the energy requirements of the total evaporation far exceeds the supply of radiant energy to the forest (Table 3.1) – at the humid tropical forest site at Janlappa, West Java, the energy requirements of the total evaporation were, within experimental error, equal to the supply of radiant energy. At the Ducke Forest site the evaporation was less than the depth equivalent of the radiation supply, but significant dry periods were recorded at this site which may have exerted soil moisture limitations.

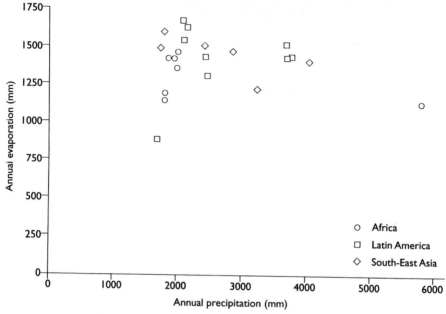

Source: Bruijnzeel (1990)

Figure 3.7 *Annual evaporation plotted against annual precipitation for tropical lowland forest*

Data compiled by Bruijnzeel (1990) for other studies of evaporation from humid, lowland tropical rainforest show a similar picture (Table 3.2, Figure 3.7). Although the supply of net radiation was not reported for these studies, it is reasonable to assume that it is in a similar range of 1500–1550 mm per year. With this assumption the results show that only in dry years, or at the Guma, Sierra Leone site, which experiences a five-month dry season, was the evaporation very much less than the net radiation supply.

As humid rain forest is able to convert annually virtually the equivalent of all the net radiation into evaporation, it is unlikely that any other land use will be able to evaporate at a higher rate. Conversion of forest to annual crops in these areas, as well as in most other areas of the world, is likely to result in increased annual streamflows.

Chapter 4

The New Ideals

THE ENVIRONMENT AND THE UNITED NATIONS CONFERENCE ON ENVIRONMENT AND DEVELOPMENT

Early Developments and the Dublin Conference

United Nations organizations have played a major role in the development of the new approach to water management. Concern over the global implications of water problems was voiced as far back as the United Nations Conference on the Human Environment in Stockholm in 1972. The UN-sponsored Conference on Water at Mar del Plata, Argentina, 1977, was seminal in the formation of the new approach and in promoting the importance of water and water management to world governments. A later water conference, the International Conference on Water and the Environment (ICWE) in Dublin, Ireland, in January 1992, saw an attendance of 500 participants, including government-designated experts from 100 countries and representatives of 80 international, intergovernmental and non-governmental organizations.

The water community represented at the Dublin conference saw it as an opportunity to put its views to the world leaders who would be assembled at the United Nations Conference on Environment and Development (UNCED) in Rio de Janeiro later in June 1992.

The water experts saw the emerging global water resources picture as critical. At its closing session, the Conference adopted the Dublin Statement and the Conference Report. The problems highlighted were not seen as speculative in nature; nor were they thought likely to affect our planet only in the distant future. They were perceived as 'here and now' and it was recognized that *'the future survival of many millions of people demands immediate and effective action'*.

The Conference participants called for fundamental new approaches to the assessment, development and management of freshwater resources, which could only be brought about through political commitment and involvement from the highest levels of government to the smallest communities. They recognized that:

'*commitment will need to be backed by substantial and immediate investments, public awareness campaigns, legislative and institutional changes, technology development, and capacity building programmes. Underlying all these must be a greater recognition of the interdependence of all peoples, and of their place in the natural world.*'

The Dublin Principles

The Dublin Conference Report sets out recommendations for action at local, national and international levels, based on four guiding principles.

Principle 1

Fresh water is a finite and vulnerable resource, essential to sustain life, development and the environment.

Since water sustains life, effective management of water resources demands a holistic approach, linking social and economic development with protection of natural ecosystems. Effective management links land and water uses across the whole of a catchment area or ground water aquifer.

Principle 2

Water development and management should be based on a participatory approach, involving users, planners and policy-makers at all levels.

The participatory approach involves raising awareness of the importance of water among policy-makers and the general public. It means that decisions are taken at the lowest appropriate level, with full public consultation and involvement of users in the planning and implementation of water projects.

Principle 3

Women play a central part in the provision, management and safeguarding of water.

This pivotal role of women as providers and users of water and guardians of the living environment has seldom been reflected in institutional arrangements for the development and management of water resources. Acceptance and implementation of this principle requires positive policies to address women's specific needs and to equip and empower women to participate at all levels in water resources programmes, including decision-making and implementation, in ways defined by them.

Principle 4

Water has an economic value in all its competing uses and should be recognized as an economic good.

Within this principle, it is vital to recognize first the basic right of all human beings to have access to clean water and sanitation at an affordable price. Past failure to recognize the economic value of water has led to wasteful and environmentally damaging uses of the resource. Managing water as an economic good is an important way of achieving efficient and equitable use, and of encouraging conservation and protection of water resources.

ECONOMICS AND STRUCTURAL ADJUSTMENT PROGRAMMES

After the momentous events in Europe following the collapse of the Berlin Wall in 1989 many of the world's major economies, including the former Soviet Union, China, India and Egypt opted for change from a centrally planned and command economy to an economy which was more market-oriented. The ideal embodied in the new economic strategy, often supported by an economic Structural Adjustment Programme (SAP), was to increase efficiency by promoting market orientation, trade liberalization, deregulation, privatization, stakeholder participation and decentralization. (Stabilization and SAPs had come into being much earlier in the 1970s as a response by the World Bank and other international development institutions to assist developing countries with the macroeconomic shocks brought about by the increase in oil prices, debt crises and world recession.)

The requirement to link SAPs with integrated water resources management is well illustrated by Zimbabwe's experiences with such a programme. In 1990, Zimbabwe, following the lead of other major economic blocs, adopted the first phase of a SAP. Here the restructuring sought to promote higher growth and to reduce poverty and unemployment by reducing fiscal and parastatal deficits; instituting prudent monetary policy; liberalizing trade policies and the foreign exchange system; carrying out domestic deregulation and by establishing a social safety net and training programmes for vulnerable groups. The focus was on the formal sector as the engine of growth. Economic SAPs are not without their opponents and even the World Bank, a major financial supporter of Zimbabwe's SAP, recognized that the performance in Zimbabwe was less than expected. Although a principal objective of the programme was to reduce poverty, it is now recognized that it was the poorest sector of the community that suffered most during the adjustment. It is also recognized that a major factor which limited the success of the first phase of the SAP in Zimbabwe was that the programme was forced through at a time when, in the early 1990s, southern Africa was suffering a severe drought. The tight interrelationship between water management and economic management is demonstrated nowhere more clearly than in southern Africa, and Zimbabwe was quick to realize that a water resource management strategy (see Chapter 5) was needed to go hand-in-hand with the SAP.

The relationship between SAPs and the natural environment is now also raising concerns although it appears very difficult to make generalizations about the impacts of these programmes. Studies carried out by UNEP

(Panayotou and Hupe, 1996) indicate that environmental impacts are very site specific and can be positive or negative.

Although there is much debate on the best mechanisms for implementing a SAP to reduce the hardships inevitably created in some sectors, there is little doubt that the principles and ideals underlying these programmes are likely to remain a basis for economic development for many countries. The recognition of the need to take into account the links between basin economics and land and water resource developments also raises questions as to how both water and land use can be valued.

The Value of Water

Enshrined in the UNCED principles is the concept that water should be recognized and treated as an economic good and economists are now able to devise methods for calculating the value of water as a commodity in its many uses. But in many non-western cultures water has value above that of a mere commodity: it has a spiritual value. To the African water is seen as 'the giver of life'; to the Maori it is seen as an 'essential ingredient of life', 'a gift handed down by their ancestors' (see Chapter 5).

Perhaps these views should not be regarded as conflicting but as complementary if, through higher valuation, whether economic or spiritual, they result in a better appreciation of the resource and more sustainable and less wasteful usage.

The advantages to be gained by ascribing a value to water and charging users a tariff related to this value are numerous. They include:

1   Provision of funds for water developments.
2   Reducing demands on the public sector for the capital and recurrent cost of providing water.
3   Releasing water for higher value use and assisting the prioritization of water allocation.
4   The resolution of resource conflicts between, for example, demands for water, power, fisheries and transport.
5   Environmental benefits, through demand management, by releasing more water for environmental usage.
6   Demand management, reducing the demand for water.

Provision of funds for water developments becomes increasingly important as cheaper sources of water become used up. The World Bank (Bhatia and Falkenmark, 1992) estimated that the cost of providing water from the 'next' project was often two to three times that of providing water from the 'current' project. Although US$10 billion was being spent each year on improving water supply and sanitation in the developing world in the early 1990s, it was estimated by the World Bank that the investment would, even if costs were fixed, have to be five times this rate if reasonable water services were to be had by all by the year 2000. Cost recovery is therefore an important feature of a water pricing policy.

Reducing demands on the public sector and donor organizations as a result of cost recovery should, in theory, release funds for other forms of development. The World Bank estimated that in 1991 only ten per cent of the cost of water projects that it funded were financed by internal cash generation.

Releasing water for higher value uses will come about if water pricing is set at such a level that it discourages lower value uses. Water pricing can therefore be an effective instrument in water allocation.

The resolution of resource conflicts may be aided by realistic resource pricing instruments. Competition between agriculture, environmental and hydroelectric requirements for water would be eased in Zimbabwe through water pricing, and agricultural demands for electric power for pumping groundwater for irrigation in India might be alleviated through realistic pricing of both water and electricity (see Chapter 5).

The environment as a valid user of water is becoming increasingly recognized. Also recognized is the increasing damage to the environment that comes through the growing supply and consumption of water. Abstractions for water supply, agriculture and industry mean less flow in rivers and the drawdown of water bodies. Lower flows in rivers also reduce the assimilative capacity and, with the discharge of pollutants, may lead to toxic conditions for wildlife and extra costs to public services for the treatment of water. Demand management, through curbing water use, may have environmental benefits. In Australia increasing charges for irrigation water is one of the instruments proposed to curb increasing salinization problems (see Chapter 5).

Arguably, within the context of IWRM, the greatest benefits to be achieved from a pricing policy for water are in relation to controlling demands for water in situations where the water resource is close to being fully exploited.

The principles of demand management and the methods that have been developed for valuing water are described below.

Demand Management

'The equivalent of the Hippocratic Oath for water engineers is to promise to meet all reasonable needs for water without question by enlarging and improving supplies.' (Winpenny, 1992)

In many developed countries the ethos of the water engineer was to equate efficiency with maximum use of the water resource for the users. Taken to its limit, any residual flows in rivers were seen as wastage. With this ethos, environmental and ecological considerations and downstream users, especially if they were trans-national users, were given little weight in the quest for 'development'.

This ethos still reigns in water departments in many countries in both the developed and developing world. Career development for water engineers is seen through association with large civil engineering projects – the larger being the better. It has been likened to a 'right of passage' for water engineers. But this ethos is no longer the powerful driving force it was of yesteryear. Water engineers are to some extent the victims of their own success. Most of the world's best sites for water developments have already been developed and the costs of developing the remaining sites are becoming prohibitive. Even where governments and donors are prepared to overlook the massive environmental and socio-economic costs of developing the remaining water resources the sheer economic costs involved are usually a sufficient disincentive. Except for 'Trade for Aid' deals, international funding for large water developments is increasingly difficult to obtain. Only in the few remaining countries which have not yet fully shaken off the constraints of command economies are large water developments still going ahead, an example being China and the Three Gorges Project. In simple terms the money for large water developments is no longer available. The trend away from supply-driven large water schemes towards greater demand management was well summarized by Arthur (1997); he recognized that '*in the background of the differentiation between need and want, lies a much wider ideological conflict about the limits to growth*'.

In most countries the accepted 'knee jerk' response to water supply shortages was to augment supplies. In the short term, demand management devices such as rationing, prohibited uses, public exhortation and the use of standpipes or the cutting-off of supplies, are widely applied. But these non-price devices, although often effective, are both costly and inconvenient to users and do not take account of the relative value of water to different consumers.

Where water is provided to users at a price less than the supply cost, a situation common in most parts of the world except the UK, the incentive for conservation and waste reduction is absent. This negative demand management leads to the paradox that in a situation where the water resource is already under stress, the subsidy is actually encouraging users to make additional demand upon it.

The alternative approach, consistent with the concepts of IWRM, is to recognize that water resources are limited and new sources cannot be developed indefinitely. Demand management in some form must ultimately be applied and the use of pricing is an instrument which can be both effective and can be defended by rational and objective arguments. If water is viewed simply as a commodity, forgetting for the moment its spiritual and aesthetic considerations in many cultures, it would be reasonable to expect that it should be priced to cover at least its cost of provision and priced so that low value uses are discouraged and supplies are available for the higher value users who are able and willing to pay for it. Clearly the strict adherence of these principles needs to be treated with some caution to ensure that, for example, the poorest sections of the community are not disadvantaged and that the application of these economic principles do not contravene the accepted spiritual and cultural beliefs associated with water. This need not necessarily be a problem. Under water subsidy conditions it is generally the richest sections of the community, which are pipeline-connected to the supply, whether for irrigation water or domestic purposes, that receive the greatest benefits in receiving water below the provision costs. It is the poor urban dweller who buys water by the bucket from a standpipe salesman who often pays the highest per-unit costs.

Although governments see many political hurdles in introducing water pricing policies – the principal concern being a short-term loss of votes and a secondary concern being the inflationary effect of charging for a commodity that is universally used – the benefits are increasingly being regarded as outweighing the disadvantages.

The development of a pricing policy for managing water demand does, however, require a methodology for deciding on the value of water (see above) to determine the price to be charged. The impacts of the pricing policy, before the policy is implemented, is another requisite and to do this, some sort of modelling study is required (see Chapter 6) to ensure that the impacts on all the affected stakeholders are understood.

Water as an Economic Commodity

Traditionally water has been regarded as a 'free' resource of unlimited supply with zero cost at the point of supply. Users have been charged for often only a proportion of the costs of transfer, treatment and disposal. Opportunity costs for water are often ignored and as a consequence users have little incentive to ensure water is used efficiently and not wasted.

The new (economic) approach to the allocation of water is to use prices and markets to ensure efficiency and that water is supplied to its most valuable uses.

In an economic sense, efficiency requires that:

1   Marginal benefit of use should be greater than the marginal cost of supply. This means that the users are able to derive greater economic benefit from the next unit of water supplied than the cost expended by the supplier in providing that extra unit of water.
2   Marginal benefit per unit of resource is equal across all uses. Where the supply of water is not unlimited and the marginal units of supply have a positive cost (that is the supplier has to expend money to develop the next unit of supply water), the value of water consumption is maximized when net marginal benefits are equal in all uses. This implies that efficiency would be increased by transferring water between users until the marginal value of the water to the users is the same across all sectors. When equality of marginal values is achieved, further redistribution of water would make no sector better off without making another sector worse off.

In theory it is possible to derive demand and supply curves which show both:

1   The marginal benefit obtained from consumption and the willingness to pay.
2   The marginal cost of supply and the willingness to supply at given prices.

Where the demand and supply curves intersect defines a theoretical price at which economic efficiency and welfare is maximized. In practice, applying market principles to water management is complicated because water does not fit the economist's model of a normal commodity and of a perfect market. Often water providers occupy a monopoly situation, the logistics of transferring water between users and sectors may be difficult, and there may be values attached to water which are difficult to quantify in monetary terms; the economist's approach may need qualification in real world situations.

Valuing Water

Various methods have been developed to determine the value of water in its many uses. Those reviewed by Winpenny (1996) include

1   Willingness to pay (WTP).
2   Marginal cost analysis.
3   'Netback'.
4   WTP, taking into account recycling and reuse costs.
5   In-stream value for pollution assimilation.
6   In-stream value for transport.
7   Hydroelectric generation with and without capital costs included.
8   Travel cost and contingent valuation methods for amenity and recreational use.

For urban household consumers, the WTP approach has often been adapted to water pricing. This involves determining the demand curve the relationship between the amount of water used by consumers and the charging price. This information can be obtained either by survey or from knowledge of the change in consumption following a price change. The total area under the supply/price or demand curve is regarded by economists as the benefit or value of the water.

For agricultural usage the value of water is often determined in terms of marginal productivity. Here the economic benefits through increased yield or quality of the crop resulting from a unit addition of water are calculated under conditions when all other farm inputs are held constant. An alternative method known as 'Netback' is related to the WTP principle. By taking the gross value of the crop per hectare and subtracting all production costs, and if required, a capital recovery cost and an acceptable profit margin, the remainder can be viewed as the maximum WTP for the water.

The same methodology can be applied to industrial water valuation, but as water is usually only a small part of total production costs, it would be misleading to attribute the whole of the remainder in the value and cost calculations to water valuation. For industrial use, another method is to regard the cost of recycling water following water treatment as the upper limit on industrial WTP. The reasoning would be that if water pricing were greater than the treatment costs, industry would opt for treatment rather than buying in water.

In-stream values for water arise through its natural capacity to assimilate waste and pollutants. This value can be compared with the alternative cost of reducing the pollutants at source or the additional costs incurred in water treatment. In-stream values for water for transport can be calculated as the cost advantage of water transport over the next cheapest form of transport (Winpenny, 1996).

Values of water for amenity and tourism have been calculated using the travel cost and contingent valuation methods. The travel cost method infers the amenity value from the travel costs of visitors to the site which are then used to construct a hypothetical demand curve for the amenity or recreational value of the water body. The contingent valuation method relies on opinion surveys to reveal the value that visitors derive from the water body.

Winpenny (1996) has reviewed the values obtained for different uses of water in different parts of the world using the methods outlined above. Although he introduces the caveat that it is important to examine the fine print associated with each valuation before considering the valuations as representative of the sector, there is a general picture that emerges: industrial uses together with speciality crops and domestic uses are always high value. Recreational uses can be high value, whilst the majority of agricultural crops, often the largest overall consumers of water, are relatively low value uses (Table 4.1)

Within many cultures, water has more significance than can be attributed solely in monetary terms. In many cultures water has an aesthetic or spiritual significance which cannot be quantified monetarily. In these circumstances the economic approach to the valuation of water needs to be tempered with other appreciations of its value within the context of IWRM.

If the Hippocratic Oath to the water engineer is to supply all reasonable needs, the 'Holy Grail' to the economist is to achieve optimal economic efficiency where marginal benefits for the resource are equalized. Within complex real world situations, the search for optimal solutions may be fruitless, but a move in the right direction, towards satisfying economic efficiency and the other objectives of IWRM, is perhaps the goal that should be striven for, and is one that ultimately is more achievable.

The Value of Land Use

Whilst methods have been devised for valuing water in its many diverse uses, including those for water supply, irrigation of crops, hydroelectricity generation, industrial production, mining, amenity and recreational uses and for the environment, there is also a need for methods for valuing land use. Within an IWRM context this is particularly true for land uses such as forests which, although having a high impact on water resources, may have valuable multiple uses. These may include not only primary uses for timber production and other forest products, but also secondary uses for recreation, conservation and tourism. The valuation of land uses for primary production is inherently straightforward although tax incentives and production subsidies may significantly affect the valuation. Valuation methods for secondary uses are less straightforward and more controversial but nevertheless important; in some

Table 4.1 *Estimated value of water for different uses (US$ per 1000 cubic metres)*

Use	USA (Gibbons, 1986)	China (Dixon et al, 1994 Kutcher et al, 1992 Adams et al, 1994)	UK (Rees et al, 1993 Bate and Dubourg, 1994)	Zimbabwe (Winpenny, 1996)	South Africa (Hassan et al, 1995)
Industrial process	180–800	500–4000		>230	
Speciality crop	100–800		230 (fruit) 1890 (potatoes)	100–150	
Domestic	20–360			>120	
Recreational	10–400				
Navigation	generally 0 but 370 in some stretches				
Intermediate value farm crops		average: 30 at critical times: 90–100			100 (sugar)
Low value farm crops	10–60	<120	80–140 (field vegetables) 10–30 (cereals & grass)		100 (dryland crops) 0 (traditional livestock rearing)
Waste assimilation	0–20				
Industrial cooling	0–10				
Sediment prevention		0–20			
Power	0–40	20			

circumstances secondary land uses may be at least as valuable. Trade-offs need then to be considered between the value of water that may be forgone under a particular land use against the sum value of the different uses of the land.

Some of the methods that have been used for assessing the amenity and recreational value of land use are identical to those used for assessing the amenity and recreational value of water. Willis and Benson (1989) have discussed the use of travel cost and contingent valuation methods for assessing the recreational value of UK forests. They also discussed other 'market related' methods including the Hedonic Price Method which links environmental assets with markets for private goods and services. The linkage of wages or house prices to environmental attributes were cited as examples that could be followed with this approach, but the practical difficulty in obtaining information on housing characteristics and sale prices of houses deterred the authors from pursuing this method for estimating the recreational value of forests.

Using the travel cost method, Willis and Benson arrived at an average recreational value of UK £1.90 for one visit to a forest owned by the UK Forestry Commission. They estimated that the wildlife attributes of the forests contributed about 38 per cent towards this value and showed that the total annual recreational benefit from the Forestry Commission's forests lay between £14 million and £45 million per year, later to be reassessed at £53 million (Benson and Willis, 1991).

Neither this approach to valuing forest recreation nor its conclusions have been accepted by all. Hummel (1992) states:

> '*Most decisions on the provisions of recreational facilities in Western Europe tend to be the result of political and technical judgements rather than economic calculations. That also applies to decisions on silviculture, at any rate in France, Germany and Switzerland, where many foresters regard the British preoccupation with calculations of net discounted revenue as simplistic and typical of a nation of shopkeepers.*'

He also pointed out that the valuation of recreational facilities does not reflect 'money in the bank'; had this been the case the operational loss of the Forestry Commission would have been more than offset by the recreational income.

Recreational and environmental benefits can accrue from other land uses in addition to forestry. Within the UK the extension of public access to the countryside is increasingly being achieved through agreements with landowners under which government agencies purchase new rights of access to land (Crabtree, 1997). The valuation of the recreational and environmental benefits of land use is then required for specific areas to assess whether inclusion within an incentive scheme merits its costs.

Aylward et al (1998) have made an analysis of the use and non-use values of different land uses on the catchment of Lake Arenal, the source area for Costa Rica's largest hydroelectric facility and irrigation scheme. This landmark study, entitled '*Economic Incentives for Watershed Protection*', takes into account both valuation, particularly non-use valuation related to watershed protection, and institutional analyses. Aware of modern research that has demonstrated that forests generally reduce water flows as compared with shorter crops (see Chapter 1), the authors state in their conclusions that

> '*by examining the issue of externalities in detail the study shows that the crucial hydrological externality (water yield) and its relative direction (positive) are contrary to that expected by previous characterizations of the problem.*'

They go on to say:

> '*while considerable variability can be expected in applying the valuation and institutional analyses to other sites and conditions,*

at a minimum this case study suggests the benefits of integrating the(se) two aspects of watershed analysis under a single framework. Additional case studies and more general theoretical work should assist in the development of a defensible consensus around rules of thumb and shortcuts in such analyses that would contribute to better policy and project formulation. Such guidance is desperately needed given the current reliance on partial analyses and outdated conventional wisdom of what constitutes watershed protection and watershed management in the humid tropics.'

For the successful integration of land and water management robust and accepted methods for valuing both water and land use will increasingly be required in the future. This is especially true where land and water are associated with environmental and recreational attributes.

SUSTAINABILITY

Raising Awareness

Tragedy of the Commons

The article entitled 'The tragedy of the commons' by Hardin (1968) is now regarded as one of the seminal documents which provoked discussion and raised awareness of issues which we now conceptualize in terms of 'sustainability' or 'sustainable development'.

Referring to the tradition of a community pasture, 'the commons', Hardin gave the following example of a society that permitted freedom of action in activities that affected common property, and how it was eventually doomed to failure.

> *'Picture a pasture open to all. It is to be expected that each herdsman will try to keep as many cattle as possible on the commons. Such an arrangement may work reasonably satisfactorily for centuries because tribal wars, poaching, and disease keep the numbers of both man and beast well below the carrying capacity of the land. Finally, however, comes the day of reckoning, that is, the day when the long desired goal of social stability becomes a reality. At that point the inherent logic of the commons remorselessly generates a tragedy... Ruin is the destination towards which all men rush, each pursuing his own best interest in a society that believes in the freedom of the commons.'*

The publication of this article in 1968, coinciding with the first pictures of earth from space, and the new notion of 'Spaceship Earth', had a powerful resonance.

The collapse of city transport in megacities, overfishing of certain fish species, overcrowding by tourists of places of natural beauty and serenity, have all been given as examples of the 'tragedy of the commons'. Land use changes resulting in adverse hydrological impacts, either in terms of quantity or quality, could equally be viewed as another example of this 'tragedy'. The salinization of much of the agricultural land of Australia and of major river systems (see Chapter 5) brought about incrementally by individual farmers' actions might be seen as a particularly pertinent example within a water resources context.

Although Hardin's article has done much to promote awareness of issues of sustainability, the general use of his narrative to predict and explain the degrading effects of increasing population pressures on natural resources has been criticized. Leach and Mearns (1996) argue that in a true commons situation, local institutions facilitate cooperation between users so that resources can be managed sustainably. The 'tragedies' only occur when the system for cooperation breaks down. There are examples in the Machakos region of Kenya (Tiffen et al, 1994[1]) where it is claimed that increasing population densities in recent years have gone hand-in-hand with environmental recovery (see Chapter 1). Fairhead and Leach (1996) have also argued that the forest 'islands' around villages in the savanna region of Guinea are not the remnants of a previous extensive forest cover degraded by an increasing rural population, but the result of the rural population actually fostering these 'islands' in a landscape which would otherwise be less woody. The implication here is that population growth has resulted in greater forest not less (see Chapter 1).

The Limits to Growth

The book entitled 'The Limits to Growth' (Meadows et al, 1972) presented a crisis scenario of the depletion of the world's resources of fossil fuels, metals, timber and fish. It is now realized that many of the assumptions on which the computer predictions were based, particularly in relation to the extent of the reserves of fossil fuels and metals, were flawed and consequently the dire forecasts that were made were mostly erroneous. Although, at that time, Meadows et al successfully furthered the awareness of sustainability issues, the subsequent obvious failure of the crisis predictions has served to undermine the legitimacy of some very real concerns.

Bruntland and Beyond

The United Nations World Commission on Environment and Development, chaired by Norwegian Prime Minister Gro Harlem Bruntland, debated and investigated current concerns of the mid-1980s about the nature of development and the long term consequences of certain development pathways. The report that was produced, entitled 'Our Common Future', but widely known as the Bruntland Report (United Nations World Commission on Environment and Development, 1987) reframed the environmental debate and laid the foundations for what was to become not only a new philosophy in our approach to development issues but also a new industry to develop the tools to implement the new philosophy.

The report provided a definition of sustainable development:

> *'Humanity has the ability to make development sustainable – to ensure that it meets the needs of the present without compromising the ability of future generations to meet their own needs.'*

Fresh water, together with food, energy, basic housing and health are recognized as basic human needs. Sustainability introduces notions of sound management of the world's resources that leave the resources in as a good a condition for the next generation as we find them today.

The report also advanced seven strategic imperatives and seven preconditions for sustainability to be achieved. The strategic imperatives advanced were that:

1   Growth be revived in the developing nations, to alleviate poverty and reduce pressure on the environment.
2   Notions of equity and non-materialistic values be included in the definition of growth.
3   Essential human needs for food, housing and energy be met whilst accepting that this will necessitate changing patterns of consumption.
4   The issue of population growth be addressed particularly through reducing the economic pressures to have children.
5   The resource base be conserved and enhanced.
6   The necessary environmental risk-management technology be developed and also made available to the developing world.
7   Ecological as well as economic factors be taken into account in decision making.

Seven preconditions for these imperatives were identified as :

1   Responsive political decision making processes.
2   Economic systems which make less resource demands.
3   Responsive social systems that maintain union by redistributing both the costs and benefits of development.
4   Production systems which can operate within ecological limits.
5   Technology developments that support energy and resource efficient solutions.
6   An international order that maintains cohesion globally.
7   Responsive, flexible, self correcting governments.

The report suggests that the shift to sustainable development must be powered by a continuous flow of wealth from industry, but recognizes that future wealth creation will need to be much less environmentally damaging, more just and more secure.

The report has been widely applauded for taking a long term and strategic, rather than piecemeal, approach to dealing with sustainability.

It has nevertheless been criticized for not taking proper account of the structural relationship between the economy and the environment. Clayton and Radcliffe (1996) argue that the report assumes that economic growth can co-occur with, or even enhance, certain types of natural capital, and that increased consumption in the developing world will therefore be possible without the environmental costs that have been associated with economic growth in the industrialized nations. They suggested that

> *'given its relative lack of structural analysis, the Bruntland Report should not be taken as a blueprint for sustainability. It is, rather, a statement of principles.'*

They go on to say that:

> *'as a general rule, the purpose of such broad statements is to stimulate progress. It is usually desirable that such statements are eventually superceded by more detailed prescriptions for change. As part of this process of development, it sometimes becomes necessary to return to and revise some of the initial principles.'*

This is probably a valid assessment of the current position. The Bruntland Report has provided the stimulus for environmental organizations, industry and development agencies to rethink their strategies and to develop the detail in the processes needed to address sustainability issues. The detailed strategies, the 'blueprints' for sustainability, are now emerging, reflecting the needs of business, conservation and development interests. The emerging blueprints, the Natural Step, the Triple Bottom Line, the Sustainable Livelihoods and the Pentagon, and Integrated Economic and Environmental Satellite Accounts, are outlined below.

The Natural Step

The concepts behind the Natural Step were originated by Dr Karl-Henrik Robert, one of Sweden's leading cancer researchers and environmentalists, and are now being developed by the Natural Step Foundation. Robert was able to persuade the King, schools, and industrial sponsors to back a report on Sweden's environmental problems and on the most critical avenues for action following the Natural Step approach. As a follow-up an educational package was sent to every household in the country, outlining the steps needed to make Swedish civilization environmentally sustainable for the long-term future.

The Natural Step philosophy is based on four principles or 'system conditions':

1   Substances from the earth's crust must not be extracted at a rate faster than their slow redeposit into the earth's crust.
2   Substances must not be produced by society faster than they can be broken down in nature or deposited into the earth's crust.
3   The physical basis for nature's productivity and diversity must not be allowed to deteriorate.
4   There must be fair and efficient use of energy and other resources to meet human needs.

The Natural Step Foundation claims that :

> *'The four system conditions provide a descriptive framework for a sustainable society. Participants on all levels – households, corporations, local authorities, nations – can systematically direct their activities to fit into this frame by requiring all secondary goals to function as natural steps in the process of achieving the four conditions of sustainability.'*

The World Business Council for Sustainable Development (WBCSD) (see Appendix 1), a coalition of 125 international companies which claim a shared commitment to the environment and to the principles of economic growth and sustainable development, endorses the Natural Step approach. It is recognized that the first three principles pose major problems for industries concerned with mineral extraction and waste management. The fourth principle they recognize is slowly gaining acceptance in the business world whereas a decade earlier many businesses would have regarded it as outside their remit. About 50 national and international companies including IKEA and Electrolux are now training their staff in the Natural Step approach. The approach is also supported by conservation organizations (IUCN, UNEP & WWF, 1991).

The Triple Bottom Line

Whereas the Natural Step focuses on the physical aspects of sustainability, the Triple Bottom Line approach recognizes explicitly the social aspects. The ideas behind the Triple Bottom Line were developed by SustainAbility, a strategic management consultancy and think-tank concerned with foresight, agenda-setting and change management. Founded in 1987, it claims to be the longest established international consultancy dedicated to promoting the business case for sustainable development.

The elements of the 'triple bottom line' are seen as representing society, the economy and the environment (Elkington, 1997). Society is seen as dependent on the economy – and the economy dependent on the global ecosystem, whose health represents the ultimate 'bottom line'. The 'bottom line' metaphor arose from the use of the term by business to represent the profit figure in a company's earning-per-share statement. To arrive at this figure accountants will, as part of standard accounting practice, collate, record and analyse a wide range of numerical data which relates to economic performance. This approach is seen as the model for social and environmental accounting, to allow the calculation of the social and environmental 'bottom lines'. The mechanisms for doing this are still being developed and are the subject of debate, but will need to embody the notions of 'social capital' and 'natural capital'.

Social Capital

Social capital is regarded in part as human capital, in the form of public health, skills and education, but also collectively, as society's health and wealth creation potential. Ismail Serageldin, the World Bank's vice president of environmentally sustainable development, recognizes that the fostering of human capital requires 'investments in education, health and nutrition'.

Natural Capital

Natural capital is often considered in two forms: 'critical natural capital' which is essential to the maintenance of life and ecosystem integrity, and 'renewable natural capital' which can be replaced, repaired or substituted. The methods for accounting for natural capital are slowly being developed.

The three bottom lines are not regarded as stable – but in constant flux, due to social, political, economic and environmental pressures, cycles and conflicts. Elkington (1997) uses the example of plate tectonics to describe the movement of the lines:

> *'Think of each bottom line as a continental plate, often moving independently from the others. As the plates move under, over or against each other, 'shear zones' emerge where the social, economic or ecological equivalents of tremors and earthquakes occur.'*

SustainAbility's perceptions of the shear-zone interactions are:

- Economic/environmental.
  In the economic/environmental shear zone, some companies already promote eco-efficiency. But there are greater challenges ahead, eg, environmental economics and accounting, shadow pricing and ecological tax reform.
- Social/environmental.
  In the social/environmental shear zone, business is working on environmental literacy and training issues, but new challenges will be sparked by, for example, environmental justice, environmental refugees, and the inter-generational equity agenda.
- Economic/social.
  In the economic/social shear zone, some companies are looking at the

social impacts of proposed investment, but bubbling under are issues like business ethics, fair trade, human and minority rights, and stakeholder capitalism.

Sustainable Livelihoods and the Pentagon

The DFID White Paper on International Development, 'Eliminating World Poverty: A Challenge for the 21st Century' (DFID, 1997) presents the concept of the stewardship of natural resources so that the needs of both present and future generations can be met. Sustainability does not rely upon a 'quick fix' solution becoming available in the future to reverse degradation. The White Paper also promotes the concept of 'sustainable livelihoods' and this, together with the management of 'the natural and physical environment' is expected to achieve the overall goal of poverty alleviation. A 'sustainable livelihood' is defined thus:

> *'A livelihood comprises the capabilities, assets (including both material and social resources) and activities required for a means of living. A livelihood is sustainable when it can cope with and recover from stresses and shocks and maintain or enhance its capabilities and assets both now and in the future, while not undermining the natural resource base.'*

Based on the work of the Institute of Development Studies (Scoones, 1998) five types of capital assets, upon which individuals build their livelihoods, are defined:

- Natural capital.
  The natural resource stocks: for example, land, water, wildlife, biodiversity and environmental resources.
- Social capital.
  The social resources (networks, membership of groups, relationships of trust, access to wider institutions of society) upon which people draw in pursuit of livelihoods.
- Human capital.
  The skills, knowledge, ability to labour and good health important to the ability to pursue different livelihood strategies.
- Physical capital.
  The basic infrastructure (transport, shelter, water, energy and communications) and the production equipment and means which enable people to pursue their livelihoods.
- Financial capital.
  The financial resources which are available to people (whether savings, supplies of credit or regular remittance or pensions) and which provide them with different livelihood options.

The DFID approach recommends plotting access on a five-axis graph, a pentagon. Carney (1998) concedes that the plotting of assets is necessarily subjective – the axes are not calibrated – but hopes that the analysis will provide a starting point for understanding how assets translate into livelihoods. However the consequences for natural resource management of applying this new 'people-first', poverty focused approach to sustainable development are not yet known. There are concerns that the poverty alleviation approaches focused at the microcathment scale may result in 'tragedy of the commons' type impacts at larger scales. Conversely, it could be argued that IWRM approaches, although aiming to achieve net economic benefits to basin inhabitants, may not be taking sufficient account of the poorest in society.

Environmental Accounting and Integrated Economic and Environmental Satellite Accounts

Environmental movements, including the IUCN, have long recognized that the System of National Accounts as defined by the UN and implemented by governments worldwide does not accurately incorporate or take account of the environment. National accounts are the economic data systems used to calculate familiar macroeconomic indicators such as gross national product (GNP), gross domestic product (GDP), savings rates, and income per capita. They are built and maintained by governments, following standard accounting practices defined largely through an international process coordinated by the UN.

Environmental accounting is seen as the mechanism by which national accounting systems can be modified to account for the economic role played by the natural environment. To increase international acceptance and implementation of environmental accounting, the IUCN launched the first phase of its Green Accounting Initiative in 1996 (IUCN, 1998).

Because of its diverse membership, IUCN regards itself as well placed to contribute to the increasing acceptance of environmental accounting. Through its neutrality in the debates over approaches to environmental accounting, it is hoped that it may help the international community move towards greater agreement on methodology, which in the long run is regarded as essential for environmental accounts to achieve their full potential as a source of information for decision-making.

While much of their work is still in the developmental stages, many countries have experimented with implementing environmental accounting methods to better understand economy-environment interactions and to test environmental strategies. Norway is one of the few countries to have institutionalized environmental accounting as a routine government activity. In the 1970s, driven by concern about resource scarcity, Norway started to maintain consistent annual data on reserves and consumption of key minerals, fisheries, and forests. Including these data in econometric planning models has helped policy decision making in regard to increasing economic activity whilst complying with international conventions on air pollution reduction.

Other countries have been active in developing environmental accounting methodologies. The US Bureau of Economic Analysis (BEA) has been responsible for developing the system of national economic accounts which are used to produce the national income and product accounts, input-output accounts and balance sheets for the US economy. Responding to the need for environmental accounting (Carson, 1994), the BEA produced a new accounting framework that covers the interactions of the economy and the environment. This framework introduces new breakdowns that are relevant to the analysis of the interactions and extends the existing definition of capital to cover natural and environmental resources. The framework takes the form of a satellite account termed the Integrated Economic and Environmental Satellite Account (IEESA), that supplements, rather than replaces, existing accounts such as GDP.

The interactions covered are those that can be related to market activities and therefore assigned market prices. Their impacts are demonstrated through effects on indicators of production, income, consumption and wealth.

The accounts have two main structural features. Firstly, natural and environmental resources are treated like productive assets. These resources are viewed in the same way as other physical assets such as structures and equipment, and are treated as part of the nation's wealth. The flow of goods and services from them can be identified and their contribution to production measured. Secondly, the accounts provide details on all the expenditure and assets that are relevant to understanding and analysing the environment-economy interaction.

It is expected that fully implemented IEESAs would allow the identification of the economic contribution of natural and environmental resources broken down by industry, by type of income, and by product.

The BEA, referring to future developments of the IEESAs, states that:

> *'The plan calls for work to extend the accounts to renewable natural resource assets, such as trees on timberland, fish stocks, and water resources. Development of these estimates will be more difficult than for mineral resources because they must be based on less refined concepts and less data.'*

And:

> *'Building on this work, the plan calls for moving on to issues associated with a broader range of environmental assets, including the economic value of the degradation of clean air and water or the value of recreational assets such as lakes and national forests. Clearly, significant advances will be required in the underlying environmental and economic data, as well as in concepts and methods, and co-operative effort with the scientific, statistical, and economic communities will be needed to produce such estimates.'* (Carson, 1994)

The outcome of this work has not yet been published. The IUCN has suggested that the strong business lobby in the USA, particularly pressure from the mining industry, has curtailed the progress the BEA was making on environmental accounting.

The Blue Revolution and Blueprints for Sustainability

These different blueprints, the Natural Step, the Triple Bottom Line, the Pentagon, and Integrated Economic and Environmental Satellite Accounts, are not mutually exclusive. Neither are they definitive. Nor are they sufficient to provide the blueprints needed by the blue revolution for addressing all land use and water resource issues, or all the sustainability questions, raised by the case studies discussed in this book.

Nevertheless, the existing blueprints will form the starting point for what is required. The Natural Step outlines principles for non-renewable resource exploitations and disposal of wastes and pollution into the environment, but says little about social issues and how to manage trade-offs. The Triple Bottom Line brings in the social dimension, but gives little guidance on how the bottom lines can be quantitatively assessed for either the social bottom line or the environmental bottom line; at present social 'accounting' is no more than social reporting. The Pentagon approach, whilst considering five forms of capital, makes no attempt to quantify them and is therefore of little value as a tool for evaluating development options. The approach used in Integrated Economic and Environmental Satellite Accounts lies closest to the requirements of the blue revolution, by providing an objective, quantitative method for accounting for economic and environmental assets; but this approach takes no account of the social dimension and social assets.

Quantitative methods of asset accounting for the social, economic and environmental bottom line are needed for the blue revolution because qualitative concepts of sustainability, although an integral part of the revolution, are not, in themselves, sufficient. The blue revolution is not just concerned with obtaining sustainable solutions, it is concerned with obtaining solutions which satisfy all relevant stakeholders representing water resource, environmental, economic and social interests, and which recognize the inevitable trade-offs, in a quantifiable form, between the interests of the different stakeholders.

IMPLEMENTING THE IDEALS

Taken together, the UNCED, SAP and Sustainability ideals have brought in a potent mix of new ideas for water management, the implications and ramifications of which, when applied, are only slowly being appreciated.

Perhaps most significant is the slow realization that the agricultural sector should not always be seen as the unquestioned priority user of water. Under centralized 'command' economies, concepts of food security and self-sufficiency ensured that the agricultural sector received priority treatment amongst water users. But agriculture is often a low value user of water and in situations of water shortage, from an economic perspective, higher value uses for supply, industry and even power generation may bring higher returns. From an economic perspective again, water diverted to the 'environment' and to sustaining wildlife and the eco-tourism industry may also be higher value uses than that of irrigation water for low value agricultural crops. There is also recognition, in the more hydrologically enlightened countries, that forestry is a high user of water and in water-deficient countries there needs to be serious consideration of whether, through the export of timber, they are really exporting their valuable and possibly more precious water resource. South Africa is perhaps the first country, through applying the results from its applied research programme into the water use of plantation forestry, to develop a water policy based not only on the well understood 'polluter pays' principle but extending this to the 'user pays' principle. The implications are that high water users such as forestry will be required to pay an 'interception levy' to compensate for the water that forests are removing from downstream users. The levy would be based on consideration of the extra water use from plantation forest as compared with the indigenous land use. In South Africa this would normally be grassland.

Agricultural and Forestry Strategies Consistent with IWRM

To satisfy ambitions of food security and self-sufficiency, agricultural research for developing countries has traditionally focused, almost exclusively, on means of increasing productivity. As productivity and water use are intimately linked, agricultural and irrigation strategies, often FAO supported, generally call for greater exploitation of any remaining undeveloped water resources to be directed towards the agricultural sector, and usually towards large commercial farming operations. The requirements of other water users are not always considered in these strategies. To support the water demands of this perceived priority user of water, government water departments have placed great emphasis on expensive civil engineering activities associated with dam construction and the construction of transmission networks to irrigation schemes. The logic of this sequence of activities was rarely questioned within a command economy and politicians could be assured of making political capital by promises of further dam construction. 'Two dams for every district' was promised by the Minister for Lands and Water Resources in Zimbabwe.

With a move from command economies to economies responding to free market forces, driven by SAP principles, the economic logic of diverting a large share of the water resources of a water-deficient country to the agricultural sector is increasingly being questioned. With increasing support for the principles enunciated by UNCED, which call for greater recognition of equity, stakeholder involvement, the environment and for treating water as an economic good, the past logic appears increasingly flawed and a new paradigm is developing. Whereas in the past it would have been heresy to doubt that agriculture should be regarded as the priority water user, these questions are now being posed in many countries. In Zimbabwe it is being asked whether, if agriculture is contributing only 16 per cent of GDP, it should be using 80 per cent of the country's water resources (see Chapter 5).

In proportional terms, the imbalance is even more extreme in Namibia where, although 43 per cent of the country's water is used for irrigation, it contributes only 3 per cent of GDP (Table 4.2).

Table 4.2 *Water use by sector and contribution to GDP in Namibia*

Sector	Water used (%)	Contribution to GDP (%)
Irrigation	43.0	3
Cattle	25.3	8
Household + other	25.3	27
Mining	3.2	16
Tourism	0.4	4
Industry + commerce	2.8	42

Source: Pallet, ed (1997)

These water allocation questions receive greater prominence in times of water scarcity. During the southern African droughts of 1992 and 1995 when water shortages in Zimbabwe led to closure of industries, severe rationing to cities and shortage of power (from hydroelectricity generation), the practice of using the scarce water resource for large-scale irrigation of relatively low value crops such as wheat was difficult to defend. Nor was the agricultural sector which was receiving water at the low, heavily subsidized 'blend price' (see Chapter 5) being encouraged to use this water efficiently. Indeed the price was so low that the returns from hydroelectricity generation, if water had just been left to flow in the northern-flowing rivers into Kariba, would have been greater than the receipts from farmers for the use of this water for irrigation. Equity considerations in relation to water allocation for agricultural use are now also on the agenda. Whereas water rights were formerly associated with land tenure and were traditionally allocated on a 'first come first served' basis, the question of need will increasingly be used as the criterion.

The UN-supported Consultative Group on International Agricultural Research (CGIAR) is now attempting to align strategies for increasing agricultural productivity within an environment of growing scarcity and competition for water through the System-Wide Initiative on Water Management (SWIM). The first paper produced under this initiative by Molden (1997) presents a conceptual framework for considering how water is used and recycled within a basin and gives a useful definition of water accounting terms. Molden also makes the important point *'that the portion of water diverted to an irrigation scheme that is not consumed, is not necessarily lost from a river basin, because much of it may be reused downstream'*. An associated paper by Seckler (1996) discusses the concept of 'dry'and 'wet' water savings. Here 'wet' savings are regarded as genuinely beneficial savings to the whole system which allow other, possibly downstream users, to make use of the savings. 'Dry' savings occur if there are no 'downstream' benefits from the savings, which might occur if the basin drained straight into the sea. Any texts such as these are surely valuable if they can introduce the ideas into the agricultural and, particularly, the agricultural irrigation community, that there are downstream users that also need water; that allowing water to flow in a river and not to use all of it for irrigated agriculture is not necessarily 'a waste', and that there may indeed be valid environmental or downstream uses for that water.

The increasing recognition of the needs of downstream users and the recognition that the environment is also a valid 'user' of water is the reason why so many countries are now developing IWRM strategies. There is a requirement now for the agricultural and forestry sectors to take full notice of these new developments which call for a more holistic appreciation of development issues, where past objectives of 'increased productivity at all costs', can no longer be justified. Research priorities now need to be reassessed in the light of this new paradigm and sectoral strategies for the agricultural and forestry sectors need to be developed which align with the UNCED and SAP concepts embodied in IWRM strategies.

Priorities in Water Allocation

Historically, rules for water allocation have developed in a piecemeal fashion where the first user of the water often gains the right of usage through 'prior appropriation'. With increasing scarcity and increasing demands from the different sectors for a share of the water, this traditional 'first come first served' principle needs to be re-examined. The question posed is how to prioritize the manifold demands for water: the demands for basic human consumption, the environment, consumption for production including agriculture, industry, power generation, and household uses (other than for basic needs), and recreational and navigational uses. The question becomes even more pressing as demand outstrips supply and no new water rights can be allocated. This situation, generally known as 'closure' of the resource (when usable outflows are fully committed), has already been reached in many North African and Middle Eastern countries and in Australia (see Chapter 5) and has generated considerable political and socio-economic tensions (Chatterton and Chatterton, 1996).

Various approaches to water allocation have been advocated, but all would agree that some form of allocation is necessary when water resources come under stress. Inevitably conflicts will develop between the different sectors demanding water; without a rational water allocation policy, there can be no means of resolution.

Top-down approaches, which give priority to some users above others, is one approac; allocation through market forces and the ability to pay is another; and a third recognizes that the transfer of 'virtual' water between different sectors may be a way of making up for resource deficits.

A priority system proposed for arid southern Africa (Pallett ed, 1997), but one equally appropriate in the wetter regions of the world, places basic consumption for human survival and sanitation as the highest priority, with demands for the environment as the second. Consumption for production purposes is third, with navigational and recreational uses last. The authors argue that placing environmental needs before those of human production may appear to be idealistic, and an impractical luxury. But they argue that it is precisely in these conditions of water scarcity and increasing demand that water resources should be well cared for, so that they can sustain development both now and into the future.

Allocation through market forces, which allows sectors which are able and willing to pay a higher price for water to obtain a higher allocation, is another approach which is discussed in the next section. Politically this approach may be difficult to apply, particularly where there are powerful lobbies, often from the agricultural sector, which wish to retain the status quo, especially where the status quo has historically meant the provision of subsidized water. Another 'market force' instrument which can allow allocation transfers between users is the transferable water entitlement (TWE). This instrument allows the right to water to be bought or sold without the necessity of selling the accompanying land and is claimed to increase efficiency by aiding the transfer of water to higher value uses such as higher value crops or industrial processes (see Chapter 5).

A further approach, and one that has been claimed to have operated successfully for the last 25 years in countries in the Middle East and North Africa, involves not the direct transfer of water between sectors but the transfer of commodities whose production requires high inputs of water (Allan, 1992, 1996). This 'virtual' transfer of water means that food deficits can be made up by buying in food, particularly cereals, from more water-abundant regions and countries which have food surpluses, thereby freeing more water in water-scarce countries for higher value uses such as industry, services and domestic supply. Allan (1996) claims that it is the water demands for agriculture and plant production, even in the domestic garden, which are the uses which cause economic crisis and social stress. He points out that about 1000 cubic metres of water are needed each year to raise our food needs, but we require only one cubic metre per year to drink and as little as 3.5 cubic metres for domestic use. However, as much as 2000 cubic metres per year are used by affluent families which have large gardens and irrigate lawns and flowerbeds in arid or semi-arid areas. By buying in food, governments can defer the painful process that would otherwise be required to manage demand and reallocate water supplies, and may avoid the immediate political stress of confronting intransigent water users with the need to improve their water use efficiency. Allan points out that there are no engineering measures that could mobilize the 20–30 billion cubic metres per year of water needed to produce the grain being imported annually into the Middle East, and that current and foreseeable technologies cannot provide water at costs which can be accommodated by crop production systems.

A corollary to the argument that water-scarce countries should consider buying in 'virtual' water in the form of cereals thereby freeing up water for higher-value uses, is that water-scarce countries should consider whether it is prudent to export relatively low-value commodities which require a high water input. A case in point is the current dilemma in South Africa with regard to its commercial forestry operations. Usually the forestry is on the hill tops and upper slopes, with fruit growing areas, industry, townships and in some cases game parks downstream. The question of whether the forestry should be depriving higher-value uses downstream is currently an emotive issue amongst the different sectors (see Chapter 5).

The approaches outlined above are not necessarily mutually exclusive. Depending on the balance of priorities within a particular country or region, an element of free market forces could be invoked with a pricing policy for water, and this could be set up subject to constraints which would protect aspects which are thought to require protection, for example, the poorest members of society, or the aesthetic and spiritual aspects of rivers and waterbodies.

In operational research terms this would mean that the model to determine the pricing policy would be set up with an objective function which might optimize basin economics subject to predefined constraints. 'Hard system' methodologies are now available to assist with this task for both water allocation (Kutcher et al, 1992; Hassan et al, 1995) and the planning of water projects using decision support systems (Jamieson, ed, 1996; Brans et al, 1986) (see Chapter 6).

UNCED AND BEYOND – NEW ORGANIZATIONS

> *'The holistic management of fresh water as a finite and vulnerable resource, and the integration of sectoral water plans and programs within the framework of national economic and social policy, are of paramount importance for actions in the 1990s and beyond.*
>
> *Integrated water resources management is based on the perception of water as an integral part of the ecosystem, a natural resource and social and economic good, whose quantity and quality determine the nature of its utilization. To this end, water resources have to be protected, taking into account the functioning of aquatic ecosystems and the perenniality of the resource, in order to satisfy and reconcile needs for water in human activities.'*
>
> Source: Agenda 21, Chapter 18, paragraphs 18.6 and 18.8, as endorsed at UNCED, Rio de Janeiro, (The Earth Summit), June 1992.

UNCED ratified the Dublin principles for the management of water. These were subscribed to by most nations and are referred to as the Dublin-Rio or UNCED principles. They have since been interpreted as:

1   Water has multiple uses and water and land must be managed in an integrated way.
2   Water should be managed at the lowest appropriate level.

3 Water allocation should take account of the interests of all who are affected.
4 Water should be recognized and treated as an economic good.

These principles are the cornerstones of IWRM and provide the basis for the blue revolution.

The international debate on IWRM has continued since the Earth Summit. The interministerial conference on drinking water supply and environmental sanitation, in Noordwijk, The Netherlands, in 1994, reinforced the UNCED concerns. More recently, the committee on natural resources of the Economic and Social Council noted that some 80 countries, comprising 40 per cent of the world's population, are already suffering from serious water shortages and that, in many cases, the scarcity of water resources has become the limiting factor to economic and social development. It also recognized that ever-increasing water pollution had become a major problem throughout the world, including coastal zones. The UN Commission on Sustainable Development, at its second session in 1994, noted that in many countries a rapid deterioration of water quality, serious water shortages and reduced availability of fresh water were severely affecting human health, ecosystems and economic development. This Commission requested that a 'comprehensive assessment of the freshwater resources of the world' be submitted at its fifth session, and to the special session of the General Assembly in 1997. This assessment was prepared by a number of UN organizations including the Department for Policy Coordination and Sustainable Development, the Department of Development Support and Management Services, FAO, UNDP, UNEP, UNESCO, UNIDO, World Bank, the WHO and the World Meteorological Organization, together with international research organizations and experts.

The report (UN Department for Policy Coordination and Sustainable Development, 1997) generally presents a rather gloomy picture of the future for the world's water resources and calls for immediate action to improve efficiency of use and to reverse degradation trends. The report notes that water use has been growing at more than twice the rate of the population increase during this century, and that already a number of regions are chronically water deficient.

> *'About one-third of the world's population lives in countries that are experiencing moderate to high water stress partly resulting from increasing demands from a growing population and human activities. By 2025, as much as two-thirds of the world population would be under stress conditions.'*

The British government's White Paper on International Development (DFID, 1997) again restates its commitment to the UNCED principles, but reinterprets them in relation to its new declared objective which focuses on world poverty reduction:

'We are supporting international efforts through the United Nations, other agencies and bilaterally to implement Key Principles for Sustainable Integrated Water Management as set out in Agenda 21 and reiterated at the Special Session of the UN General Assembly in June 1997.

We will:

- *treat water as both a social and economic good*
- *increase our support for programmes that bring clean, safe water to poor people*
- *encourage all those who have an interest in its allocation and use, particularly women*
- *be involved in decision making and management of water resources*
- *adopt a comprehensive framework that takes account of impacts of water use on all aspects of social and economic development.'*

The allusion to 'a comprehensive framework that takes account of impacts of water use on all aspects of social and economic development' brings in concepts of IWRM. Some observers are critical that there is insufficient commitment to an integrated approach. Boyd (1997) of the ODI states

'The White Paper places strong emphasis on environmental issues, and links environmental conservation with poverty reduction. However, improving rural livelihoods demands an integrated approach to natural resource and environmental management, rather than this sectoral approach which reflects conventional disciplinary specialisations.'

The UNCED ideals and the blue revolution are leading to the formation of new international and national organizations which explicitly or implicitly accept these new directions. The international debate on IWRM has been furthered by two new non-governmental organizations (NGOs): the Global Water Partnership (GWP) and the World Water Council (WWC). Both these organizations were founded in 1996 and have similar objectives although the GWP regards itself more as a facilitator of IWRM projects whilst the WWC promotes itself as 'think- tank' on IWRM issues. The English and Welsh Environment Agency, formed in 1996, is an example of a national organization which was formed with the explicit intention of taking a more integrated approach to water and environmental management and regulation.

The Global Water Partnership

The GWP, whose proclaimed ambition is to translate the Dublin-Rio principles into practice, was formally established at a founding meeting in Stockholm in August 1996. It is promoted as an international network open to all involved in water resources management, including governments of developing as well as developed countries, UN agencies, multilateral banks, professional associations, research organizations, the private sector and NGOs. The GWP intends to emphasize its comparative advantages as compared to bilateral and multilateral donors, and other agencies concerned with water and development. It claims no wish to duplicate already existing activities or compete with established donor agencies. Rather it wishes to support those already working in the field of water and development and thus gain from the experience and knowledge that already exists. The specific objectives of the GWP are that it will

> *'support integrated water resources management programmes by collaboration, at their request, with governments and existing networks and by forging new collaborative arrangements, encourage governments, aid agencies and other stakeholders to adopt consistent, mutually complementary policies and programmes, build mechanisms for sharing information and experiences, develop innovative and effective solutions to problems common to integrated water resources management, suggest practical policies and good practices based on those solutions, and help match needs to available resources.'*

The GWP is made up of four components:

1   The Consultative Group (CG).
2   The Technical Advisory Committee (TAC), consisting of professionals and scientists in disciplines related to water resources management.
3   The Steering Committee.
4   The GWP Secretariat, based in Stockholm (see Appendix 1), which provides administrative support to the CG and the TAC.

The World Water Council

The World Water Council (see Appendix 1) was formed in June 1996, with the proclaimed objectives of 'promoting awareness of global water issues and facilitating conservation, protection, development, planning and management of world water resources'.

The WWC believes there is a need for another water organization because:

> *'Our understanding of global water policy issues suffers from the extreme fragmentation of the field into national, regional and local water authorities and a host of professional and scientific organizations established along sectoral lines. Until now there has been no overarching umbrella group able to deal with water policy issues in their entirety, identify problem areas and advocate solutions.'*

The WWC calls itself

> *'a true umbrella organization, a forum and a think-tank where
> all the divergent interests concerned with water issues can meet,
> debate, reflect and provide solutions to global water policy issues.'*

The Council is currently governed by a board of governors presided over by a chairman. The elected board of governors, whose members come from all geographic regions of the globe and from all sectors concerned with water management and policy issues, is chosen by a general assembly composed of all voting members. The Council is launching two long-term projects. The first is a long-term vision for water, life and the environment, a process that will lead to the systematic identification of the water-related problems facing the world in the next century and recommend solutions. The other is a global water assessment, an effort to develop for the first time a satisfactory inventory of the world's fresh water resources.

Time will tell whether IWRM, which already has to cope with many fragmented sectoral organizations involved with water management, can really afford the luxury of two international umbrella organizations to further its interests.

The English and Welsh Environment Agency – a National Initiative

The Environment Agency was formed in April 1996 from former regulatory authorities whose remits separately included water, pollution and waste disposal. These were the National Rivers Authority (NRA), Her Majesty's Inspectorate of Pollution, and the Waste Regulation Authorities, together with smaller units from the Department of the Environment. The stated aim of the agency is to 'protect or enhance the environment as a whole, in order to play its part in attaining the objective of sustainable development'.

The objectives of the agency set by Ministers are:

1   An integrated approach to environmental protection and enhancement.
2   Considering the impact of all activities and natural resources.
3   Delivery of environmental goals without imposing disproportionate costs on industry or society as a whole.
4   Clear and effective procedures for serving its customers, including the development of single points of contact with the Agency.
5   High professional standards, using the best possible information and analytical methods.
6   Organization of its own activities to reflect good environmental and management practice, and provision of value for money for those who pay its charges, and for taxpayers as a whole.
7   Provision of clear and readily available advice and information on its work.

8   Development of a close and responsive relationship with the public, including local authorities, other representatives of local communities and regulated organizations.

Sherriff (1996) states

> 'The creation of the Environment Agency has further strengthened the integrated approach to river basin management and has specific duties associated with sustainable development and the need to take costs and benefits into account in exercising its functions.'

in other words a clear restatement of the objectives of IWRM and the blue revolution.

One task of the Environment Agency is to consider the adequacy of water resources. It does this in a number of ways.

- It must balance the medium-term demands for, and supply of, fresh water (see Figure 4.1). The balance between supply and demand varies from year to year. There was below average rainfall in 1995 and, notwithstanding a previous wet winter, a number of UK water companies had difficulties in supplying their customers. Rainfall during the winter of 1995/96 was also well below average.
- It is required to take action, as and when it considers necessary, in order to conserve, re-distribute, or otherwise augment water resources in England and Wales, and to secure the proper use of water resources.
- It has to manage the rate of abstraction of water in relation to the average effective rainfall ('available' rainfall). Figure 4.2 shows that this varies spatially across the country at any one time. In the Thames Region water is recycled to the extent that the abstraction in 1994 exceeded the effective rainfall.
- The Agency must also make assessments of the forecast balance between future demands and available water resources. It also prepares plans aimed at achieving a proper future balance by considering not only engineering options designed to increase supply, such as new reservoirs or transfer schemes, but also demand management options, such as leakage control and water metering.
- The Agency is also required to address the issue of the impacts of abstractions on low river flows and of the environmental consequences of these low flows. A priority list of 'low flow' rivers is shown in Figure 4.3. The Agency recognizes that most of the problems have been caused by 'licences of right' which were granted under the earlier Water Resources Act of 1963. These authorizations legalized existing abstractions without reference to their environmental impact. The Agency is seeking to progressively resolve these problems with the cooperation of the relevant abstractors, most of which are water companies.

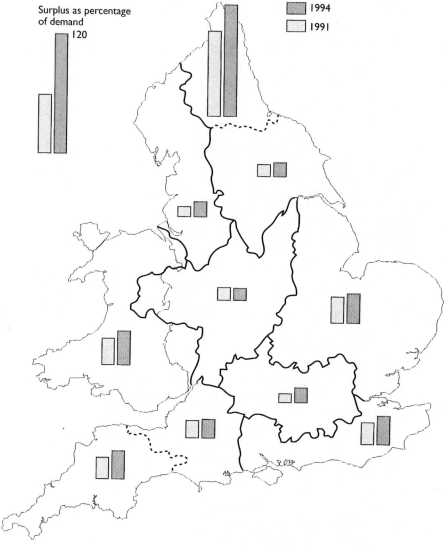

Note: Average demand is the distribution input, plus non-potable supplies to industry.

Source: Environment Agency and NRA

Figure 4.1 *Public water supply surplus of the total available water resource yield, over the average demand for 1991 and 1994*

Sherriff (1996) remarked that the recent droughts of 1995/96 had resulted in a shift of emphasis from water quality to water quantity as being the top priority for water companies' future investment plans. The impacts of future climate change on water resources (Arnell, 1996) and of land use change, especially that resulting from the proposed doubling of lowland forests in the UK (see Chapter 5) are also important issues which may result in supply shortages in times of increasing future demands.

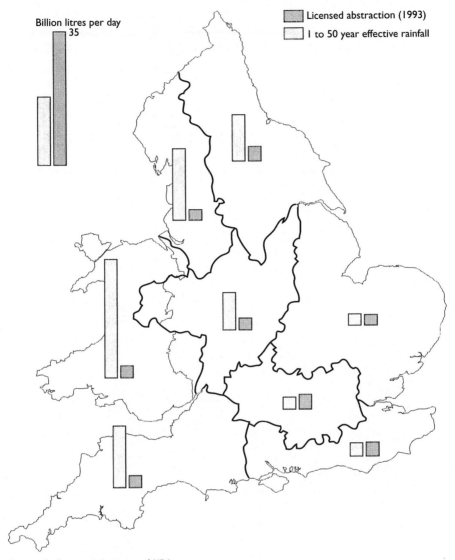

Source: Environment Agency and NRA

Figure 4.2 *Rate of abstraction of water in 1994 in relation to the average effective ('available') rainfall for different regions of England and Wales*

The Environment Agency (EA) operates over eight regions which are further subdivided for administrative purposes, allowing an integrated approach to be taken at a local level. Following on from the work of the NRA, the river catchment (watershed) has been adopted as the spatial unit within which environmental decisions are made and implemented.

Source: Environment Agency and NRA

Figure 4.3 *Location of the priority 'top low-flow' rivers where environmental and ecological damage is occurring*

Local Environment Agency Plans

Local Environment Agency Plans (LEAPs) are non-statutory action plans which use the local planning process to promote inputs from stakeholders such as local authorities, industry, farming organizations, environmental groups and other agencies and individuals, including local residents. They play a central role in developing liaison between the EA and the other stakeholders, educating the public on local issues, prioritizing local environmental issues and in providing the mechanism through which they can be resolved.

The LEAP process evolved from that developed by the former National Rivers Authority for producing Catchment Management Plans (CMP). It was recognized (Slater et al, 1995) that the production of many of the earlier CMPs had involved little interaction with stakeholders and that, partly as a result, the CMPs had little impact as planning tools. An exception was the Thames region of the Authority, which had developed a more systematic approach to liaison with local planning authorities and other stakeholders and it was the ethos and the procedures developed by Thames region which largely defined the subsequent LEAPs process (Gardiner, 1992)

The production and implementation of a LEAP involves three stages:

1   Following consultation with the relevant stakeholders and analysis of the issues affecting each catchment, and an assessment of catchment uses and resources, a consultation report is produced. This then forms the basis for public consultation and discussion.
2   Following public consultation the EA then produces an action plan which details areas of work and proposed investments together with timescales, targets and estimated costs for improving environmental conditions within the catchment.
3   The implementation of the LEAP goes hand-in-hand with appraisal and monitoring and an annual review is produced which reports on progress.

The gap between the ideals embodied in the LEAPs and their production, in a consistent form, by the local branches of the EA may need filling through a process of self-education and the provision of both technological and soft system tools. These aspects are discussed in Chapter 6.

Chapter 5

Water Resource Conflicts

There are many causes of land use and water resource conflicts. They arise because of land use demands ranging from the most fundamental – for land uses to provide both basic food supplies and water of sufficient quantity and quality to satisfy basic human needs for drinking and sanitation – to the more aesthetic, for example the desire for having attractive land uses such as forests and water bodies as visual enhancements to the landscape. Tensions between the many different users –agriculture, forestry, industries, power, mines; urban and rural consumers; amenity, ecology and environment – exist in many parts of the world. Rights of access to water and equity considerations are highly emotive political considerations as is the question, in some countries, of how water can be shared across international river basins. Consideration of the sustainability of the land and water resource for both present and future generations, always of paramount concern in some cultures, is now more appreciated by the western world. Although the land use and water resource issues and concerns are often as diverse as the different countries' cultural, economic and technical development, there is nevertheless a certain common approach as to how governments treat the issues. This commonality has manifested itself as a steadfast reluctance to deal with the issues, particularly those that are not attractive from the point of view of gaining short-term political capital: to defer and fudge the issues as long as possible, has been the norm. Lack of awareness and lack of research may sometimes have been mitigating factors, but even where land use and water resource issues have been well researched and understood, the vision and enthusiasm for facing and taking up the issues is often lacking. Increasing pressures on governments, both external and from within, are now becoming such that procrastination is no longer an option. In some more enlightened countries, the revolutionary tide has turned.

Some of the major resource conflicts in the Nile basin, Zimbabwe, Malawi, India, Australia, USA, UK, New Zealand, the Philippines and South Africa, and how these relate to the blue revolution, are outlined below.

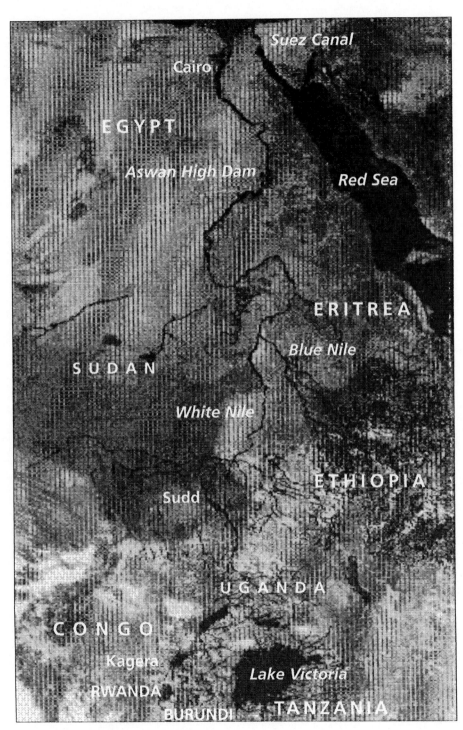

Source: TECCONILE

Plate 5.1 *The Nile basin*

THE NILE – A TRADITION OF POLITICAL CONFLICTS

The modern history of the exploitation of the Nile waters demonstrates the past ascendancy of the traditional, single-minded, engineering approach to river and water resources management. It also demonstrates how this approach has led to significant economic benefits for the riparian countries of the Nile, particularly Egypt. History also shows a tradition of political conflict between many of the states exacerbated, if not caused, by these engineering developments. More recently the environmental and socio-economic impacts of these developments have been questioned and it is these concerns which are likely to moderate, if not halt, future large-scale developments.

Nine African states share the Nile catchment which covers one-tenth of Africa and provides water to the world's longest river. Together with the political boundaries, the catchment has within it well defined climatic, social and religious divides stretching from the desert in the north, with an Islamic tradition, to the well-watered mountains of Ethiopia in the south, the home of a mostly Christian population.

An authoritative history of the early engineering developments on the Nile and the involvement of British engineers and hydrologists who brought them about, has been given by Collins (1990). Newson (1992a, 1997) supplements this history with insights into the mechanisms and reasons behind the various basin development schemes.

Engineering Developments and Water Sharing

The first phase of engineering developments, in modern history, began with the construction of the dam at Aswan in 1902. Claims on Nile water by Egypt had been established through long historic use. But increasing use of the upstream waters of the Nile by Sudan for irrigation, in the early part of the 20th century, necessitated some form of agreement over the sharing of the waters between the two states, and this led to the Nile Waters Agreement of 1929. The agreement served only to partition flows between the two states; ideas for the development of the basin as a whole and for increasing its storage came later. Engineers saw the Sudd – the wetland region in southern Sudan with floating vegetation which detains and evaporates about half the Nile flows – as a major obstacle to developing the water resources of the basin. In 1925, the Egyptian government approved a scheme to 'canalize' the Nile through the Sudd to reduce evaporation losses, but the Sudan government was not adequately consulted or involved in the decision-making process and the proposals came to nothing. By 1938 a number of ways of channelling the Sudd had been evaluated and the Jonglei Canal Diversion Scheme became the scheme preferred by Egypt. Again through lack of consultation with the British in Sudan, the scheme did not progress (Collins, 1990).

The second phase of major engineering – involving the construction of the Aswan High Dam a little upstream, and within the reservoir of the original dam built in 1902 – became a source of major international conflict. Refusal by the USA in 1956 to fund the construction of the dam led President Nasser to nationalize the Suez Canal as a means of generating internal funds. Later the former Soviet Union funded the construction of this, the world's largest dam with 180 watergates and 12 power-generating units supplying 2.1 million kw of electric power.

Conflict and Conflict Management

The competition over access to the Nile's waters by the riparian states continues. Upper riparian countries are starting to develop their water resources to meet growing population needs. Since Egypt would be particularly affected by additional water withdrawals in the upper basin, growing tensions exist between Egypt and these upstream states. In particular, Ethiopia's plan to divert the Blue Nile's water for irrigation projects causes particular concern to Egypt as this tributary supplies 80 per cent of the Nile water entering Egypt.

Treaties, basin organizations and commissions and 'initiatives' are all now playing a part in conflict management. In 1959 Egypt and Sudan established a treaty on water allocation whereby Sudan receives 18.5 and Egypt 55 $km^3.yr^{-1}$. Riparian countries of the Kagera, an important tributary of the Nile, created a basin organization in 1977, and Tanzania, Rwanda, Burundi, and Uganda are now part of it. Egypt has also unsuccessfully attempted to create a basin-wide organization, the failure of which was seen as being in part due to civil wars in Ethiopia, Sudan, Rwanda, and Burundi and also to political tensions between Egypt and Sudan. In 1992 the 'Nile Basin Initiative' was launched by the Council of Ministers of Water Affairs of the Nile (COM) basin states to promote cooperation and development in the basin. Six of the riparian countries – the Democratic Republic of Congo, Egypt, Rwanda, Sudan, Tanzania, and Uganda – formed the Technical Cooperation Committee for the Promotion of the Development and Environmental Protection of the Nile Basin (TECCONILE) (See Appendix). One of the first activities of TECCONILE was the preparation of an action plan and after two workshops, one in Entebbe, Uganda in June 1994 and the second in Cairo, Egypt, in November 1994, the 'Nile River Basin Action Plan' (NRBAP) was developed which defined 22 development projects covering integrated water resources planning and management, capacity building, training, regional cooperation, and environmental protection and enhancement. The organization holds the promise of not only resolving water conflicts on the Nile, but also achieving true integrated water resources management for the basin. However, the report of the sixth meeting of the COM held in Arusha, Tanzania in March 1998 shows that although there have been many achievements, there is still a long way to go before that goal can be reached. The report makes the following observations:

1   The Nile River constitutes key natural resources in our respective countries and whose potential largely remain underdeveloped in most of

the countries. There are emerging conflicts over access and use of the resource among the countries.

2   Five out of the ten poorest or least developed countries in the world are found in the Nile Basin.

3   Strengthened cooperation and collaboration is needed to remove the current impediments including inequity in the access and use of the resource and set a framework and mechanism to promote equitable allocation and use of Nile waters for socio-economic development of the riparian countries.

The report also recognizes that *'as time passes by we must find a solution to managing and sharing the resources of our precious basin'* and *'a belief that a shared vision can only be legitimised by action on the ground, action that benefits the peoples and particularly the poor and the disadvantaged in the Nile basin.'*

Trade-offs: Economics, the Environment, Society and Sustainability

The water resource developments on the Nile illustrate the complexity of the issues and the temporal and spatial trade-offs that need to be considered within IWRM.

On the one hand, engineering developments provide irrigation water, hydroelectricity, regulated flows and the avoidance of damaging and life-threatening floods, improved navigation for river transport and, perhaps most importantly, the potential through irrigation for feeding growing populations who can benefit from the developments. The Aswan High Dam development also provided the electric power for supplying chemical fertilizer and steel plants which allowed for industrialization in Egypt and Sudan.

On the other hand, engineering developments may be introducing a number of unforeseen problems with different spatial and temporal dimensions. Increased irrigation from the Nile waters following the construction of the dams is resulting in rising groundwater levels, increased salinity and water-logging which, although providing economic benefits for the present generation, may be leaving possibly insurmountable environmental rehabilitation problems for the next. The reduction of the natural deposition of silt on the downstream floodplains now has to be compensated for by the addition of some 13,000 tonnes of lime-nitrate fertilizer (Loucks and Gladwell, 1999). The reduction in the sediment load carried by the Nile is resulting in coastal erosion, as had been predicted. Fish catches have been reduced to about half those before the Aswan High Dam was built (ibid). Although major international efforts resulted in the translocation of some historic monuments, others were lost under the reservoir. The loss of land under the reservoir also forced the translocation of some 100,000 Egyptian and Sudanese Nubians.

It is now realized that the engineering developments planned for the Sudd would not have been without environmental and social costs. The present economy of the area is dependent on the natural environment and the grazing provided by the river through the annual cycle of river flow and the flooding of the floodplains (Sutcliffe, 1974). Any changes in the relation of the river to the bank and floodplain would be expected to have an exaggerated effect on the vegetation. The water evaporated in the Sudd cannot be regarded as a total loss (Howell et al,1988) as it is an essential component of the local grazing and fishing economy of the Jonglei area. Benefits to downstream users would have been at the expense of the communities living in the Sudd who have made use of the water from time immemorial and which is theirs by right.

ZIMBABWE AND ITS WATER RESOURCES MANAGEMENT STRATEGY

The government that came to power in Zimbabwe at independence in 1980 invested heavily in health and education and, through parastatal organizations, fostered rural development and the productive sectors in an attempt to reduce socio-economic disparities. This led to an increase in public expenditure, which for most of the 1980s amounted to 45 per cent of the GDP. Although social indicators improved, particularly in health and education, per capita income stagnated. Large government spending reduced incentives for private investment and fuelled inflation, while shortages of imported goods constrained investment and growth. Population grew faster than job creation, widening the disparities in income levels. In 1991, the government proposed a policy agenda that formed the basis for the Structural Adjustment Programme (SAP). The World Bank supported this with a US$125 million structural adjustment loan (SAL) and a US$50 million structural adjustment credit (SAC), both approved in 1992 and finalized in 1993.

The beginning of the adjustment programme coincided with the severe southern African drought of 1992 and left Zimbabwe in its worst recession since independence. The drought badly affected the implementation of the SAP but it also brought to the attention of all Zimbabweans, in the most dramatic fashion, not only the crucial importance of water in a country which is subject to frequent droughts, but also the desperate need to have a better way of managing the country's water resource.

Traditionally, water-resource developments in Zimbabwe had been geared towards providing water to the commercial agricultural sector through the construction of impoundment reservoirs and irrigation canals. Currently annual surface water usage averages 4.7×10^9 m^3 out of an estimated potential of 8.5×10^9 m^3. Agriculture accounts for approximately 80 per cent of this usage and it is generally the large commercial agricultural sector that has been most vocal in calling for the remaining 3.8×10^9 m^3 to be brought into production, even though it generally recognizes that the easiest and least costly sites have already been used. The focus on the agricultural sector was in response to the desire, both before and after independence, for the country to be self-sufficient in food production.

During the droughts agriculture was not the only sector to be affected. Mining and manufacturing industries suffered restrictions and sometimes closures. Supplies to the major cities and urban areas were restricted and hydroelectric generation was also reduced. There was also a greater recognition that although the agricultural sector was important to the economy, contributing 16 per cent of GDP, this was at the expense of it being the major water consumer. Questions about the balance of priorities were raised by some city dwellers in Harare, suffering water restrictions, who, during the drought of 1995, saw large commercial farmers irrigating wheat (a low value crop) on the outskirts of Harare. By contrast, in one of the provincial centres, Mutare, the commercial farmers complained that their livelihoods were put at risk when, during the same drought, as an emergency measure and after due process, abstraction licenses were revoked without compensation, and water which would otherwise have been used for irrigating crops was used to supply the city.

Other questions raised were whether the agricultural sector, if only providing 16 per cent of the GNP, should be the most privileged user of water and whether the price it was paying for the water was too low. Bearing in mind that in the early 1990s the agricultural sector was paying for water at the 'blend price' of 35 Zimbabwe dollars per thousand cubic metres – a value less than the value of the power generated if it had just been left to flow in the northern-flowing rivers into Kariba – this could be considered a fair question!

The competing demands from the different sectors for Zimbabwe's water, highlighted so dramatically during the droughts, led to other water-related conflict. Increasingly questions of equity, the right for all people to have equal access to Zimbabwe's waters irrespective of whether they had land tenure, became an issue. It was also appreciated that the environment was a valid user of water and that other downstream users and nations also had rights to the water.

The single-minded drive by the agricultural sector to develop the remaining unexploited resource for agricultural production, as called for in Zimbabwe's FAO supported irrigation strategy, could be seen against the downside of increased environmental damage as evidenced by more 'dry' rivers, such as the Save, and less water for the downstream nations of Mozambique and South Africa.

These were some of the tensions which led to general recognition throughout virtually all sectors and at all levels in society in Zimbabwe that something had to be done quickly to redress the balance, to properly take into account the equity, environmental, economic and trans-national issues.

Zimbabwe has now adopted a programme for the development of a National Water Resources Management Strategy (den Tuinder et al, 1995; Calder, 1997), supported by four national donor organizations, GTZ, DGIS, NORAD and DFID. The programme has a wide remit to develop a new water act and the structure for the proposed new Zimbabwe National Water Authority (ZINWA) and to develop a strategy consistent with UNCED and SAP principles.

MALAWI: LAND USE CHANGE AND LAKE LEVELS

Malawi, a country with a fifth of its area occupied by Lake Malawi, one of the deepest and the fifth-largest lake in Africa, usually regards and presents itself as a country well endowed with water resources. If it were true, it would be fortunate, as the economy of the country is heavily reliant on its water resources and the lake. Virtually all the country's electricity is generated from hydroelectric schemes on the Shire River which drains the lake. The fishing industry on the lake provides most of the country's protein food supply. The tourism industry is centred on the lakeside hotels. The ecology of the lake is unique, containing many hundreds of species of cyclid fish and ranks high in the UNESCO classification of World Heritage sites. The lake also provides an important national and international network for water-borne transport. But sadly the sanguine view presented of a country with secure and abundant water resources is a chimera.

Although the functioning of the lake is so critical to the viability of Malawi, until recently little was known about the reasons for the large changes in lake level that have occurred over the last 100 years for which records exist. For example, during the early 20th century there was no outflow from the lake and it became an inland drainage system. Various explanations have been put forward to account for these changes including tectonic movement, blockage of the lake outlet and linkages with sunspot cycles.

To address this issue, a team of researchers from the UK and hydrologists from the Water Department of the Ministry of Works in Lilongwe carried out a water-balance modelling study of the lake using the historic rainfall and lake-level data. The study was carried out to determine the cause of these changes and to determine to what extent these changes in level might have been the result of land use changes, the most significant of which has been the clearance of the dry deciduous miombo woodland for rainfed agriculture.

The evaporation model which was developed (Calder et al, 1995) considered the catchment to be composed of one of three surface types – forest, agricultural land or water. Values for the parameters for the model were derived from previous land-use evaporation studies carried out in India.

Using the recorded rainfall data for the last 100 years, and by applying the present relationship between lake level and the outflow from the lake (stage/discharge relationship) for the whole period, the model was set up to generate predicted lake levels. With a value of 64 per cent for the forest coverage of the catchment, this model was able to describe well most of the major fluctuations in level during the period from 1896 to 1967, both seasonally and annually (Figure 5.1). (An exception was a period in the 1930s when the stage/discharge relationship for the lake may have been affected following the prolonged period of no outflow in the early part of the century.)

The overall agreement between prediction and observation indicated that variations in rainfall alone, without invoking any other changes in either evaporative demand or the hydraulic regime of the lake or other esoteric explanations such as sunspot cycles, was sufficient to explain lake level changes during this period.

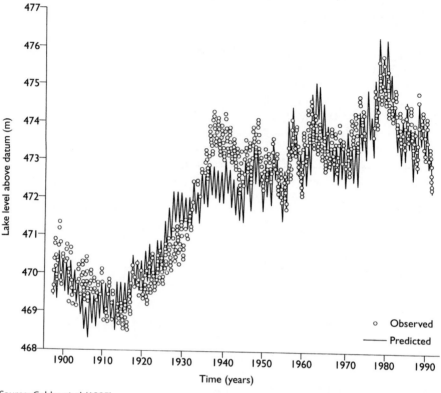

Source: Calder et al (1995)

Figure 5.1 *Lake Malawi levels: observed and predicted*

For the more recent period, model predictions which take into account a decrease in forest cover of 13 per cent over the period 1967 to 1990 (consistent with measurements of the decrease in forest cover for this period) agree well with observations both annually and seasonally. Without this decrease in forest cover it is predicted that the lake level would have been almost one metre lower than that actually observed before the onset of the southern African drought of 1992. As the country is reliant on the lake for hydroelectric generation, fisheries, tourism and transport, any further lowering of the lake level would have caused much more serious disruption, implying a significant benefit to water resources of the removal of the miombo woodland.

The model has also been used, in association with the development of a Water Resources Management Strategy for Malawi, to investigate how proposed irrigation developments would affect the water balance of the lake and lake levels (Calder and Bastable, 1995).

The modelling methodology used for Malawi has been extended through the use of Geographic Information System (GIS) methods to allow the convolution of rainfall and climate patterns with patterns of land use in Sri Lanka, the lowlands of the UK (Calder et al, 1997b) and in the Lake Malawi and Zambezi basins (Price et al, 1998) (Figures 5.2 and 5.3).

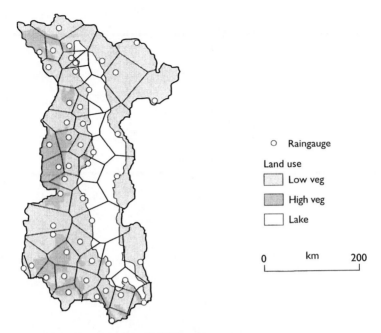

Note: used with the GIS version of the HYLUC model.

Figure 5.2 *Location of rain gauges and vegetation types on the Lake Malawi catchment*

The Zambezi basin study shows the spatial variability in the effects of deforestation on runoff (Table 5.1). The change in mean annual runoff expected as a result of a one per cent reduction in the natural forest cover of the major subcatchments of the Zambezi ranges from zero for the Rioc (where mean annual runoff is zero) to 3.33 mm for Malawi.

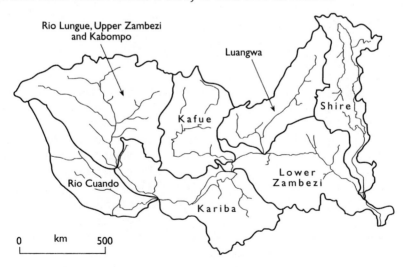

Figure 5.3 *Zambezi catchment and major sub-catchments*

Table 5.1 *Average change in mean annual runoff expected as a result of clearance of the natural forest for rainfed agriculture on the major subcatchments of the Zambezi*

Subcatchments of the Zambezi	Catchment Area (km²)	Increase in Mean Annual Runoff for a 1% reduction in natural forest cover (mm)
Lungwé Bungo, Kabombo and tributaries		2.87
Chobe	133,593	0.00
Kariba (tributaries between Victoria Falls and Kariba Dam)	223,364	0.74
Kafue	157,638	2.66
Luangwa	149,438	2.86
Shire	157,231	3.33
Lower Zambezi	237,393	2.05

Source: Price et al (1998)

These GIS developments should provide not only more accurate predictions of the hydrological impacts of land use change, but also a more general modelling approach for investigating forest/land use impact problems worldwide.

A general feature of these studies is that the conversion of indigenous forest to rainfed agriculture, whether in the wet or dry climatic regions of the world, is likely to result in an increase in annual runoff. Recognition of the impacts of land use change is therefore important in both assessing and managing freshwater resources, whether at the local or regional scale.

INDIA: EUCALYPTUS, IRRIGATION, POWER AND WATER RESOURCES

The complexity and interrelations of the water resources, and social, economic and political dimensions related to land use, reach extreme proportions in India. The tensions and passions generated by the issues are manifest at all scales. Without the strong framework of India's federal government the conflicts between the states of Karnataka and Tamil Nadu over the sharing of the waters of the Cauvery River may well have escalated. Groundwater tables are dropping in most Indian states due mostly to increased groundwater abstractions. The largest volume of the abstraction is directed to the agricultural sector for irrigation. To provide the power for pumping the water from the ground to the surface, up to two-thirds of all the electric power generated in some southern Indian states is used for this purpose alone. These rates of abstraction are clearly unsustainable and may be causing irreversible damage to some aquifers where, through compaction, their physical properties are being permanently altered. Yet at present there are neither legal nor economic controls nor the political will to address the situation. There are neither licensing arrangements for borehole drilling nor requirements for obtaining a water right for abstracting groundwater. Economic controls are minimal as the farmers receive the electric power at heavily subsidized rates and there are also many cases of illegal connections and farmers receiving power without any payment. The political will to address the situation is largely absent as politicians rely on the farmers' vote for their support. The situation is in desperate need of the blue revolution, an integrated approach to resource management, but the two triggers for the revolution, an official recognition of the problem and the will to tackle it, are not in place.

The *Eucalyptus* Concern

Other land use and water resource issues have been well recognized in India, one of the 'hottest', not only in India but worldwide, being related to the hydrological impacts of the *Eucalyptus* species. Approximately half of all plantation forestry in the tropics and sub-tropics is composed of *Eucalyptus* species. Their high growth rates and their ability to grow within a wide range of site conditions make them attractive species for both commercial and social forestry applications.

The large-scale planting of such exotic species has aroused deep-seated anxieties in India and in many other tropical countries. Eucalyptus plantations were thought to cause serious socio-economic problems at the village level and adverse environmental impacts, particularly in relation to high water use, erosion and nutrient depletion. They were often presented as the principal culprit for the lowering of groundwater tables. The uncertainty and the worries that existed are illustrated in 'The Tree that Caused a Riot' (*New Scientist*, 18 February 1988).

In the absence of hard evidence to the contrary, speculation by the press and by some local environmental groups raised the controversy to such a pitch that in parts of Karnataka State in southern India, farmers ripped out eucalyptus seedlings from government nurseries and plantations. Yet other farmers in Karnataka saw eucalyptus as a valuable source of income and were keen to plant them on their fields. And eucalyptus trees can have other benefits. As providers of a fast-growing source of timber, firewood and pulp they can help to reduce the pressure on the few remaining indigenous forests as wood sources and thus aid conservation efforts. Through saving foreign exchange on the importation of pulp they have obvious economic benefits.

Clearly information was required on the potential hydrological impacts so that decisions could be made in the context of IWRM. As a result of a comprehensive research programme and detailed studies in India, South Africa, South America and Australia hard evidence now exists on the hydrological impacts of these plantations. The results do not show *Eucalyptus* species to be quite as villainous as they have often been portrayed, but neither do they show them to be without hydrological disbenefits. What they do

Source: photo IR Calder

Plate 5.2 *Bark stripping after felling eucalyptus plantations*

show is a complex pattern of interactions, some of which may be seen as beneficial and others as adverse. Although the research providers have done their work and major advances in knowledge have been achieved, the use of this knowledge for land management purposes has been less spectacular. This is perhaps unsurprising as the issues are complex and the tools for holistic resource management such as 'decision support systems' (see Chapter 6) have only recently been developed.

Eucalyptus species are beneficial in having high plot water use efficiencies. They probably produce more biomass per unit of water evaporated on a plot basis than any other tree species, their high growth rates making them very attractive plantation species and in some countries, such as Australia, their ability to reduce and hold down water tables is a very great advantage in salinity control. Offsetting these advantages is the high water consumption that needs to be considered together with other socio-economic and environmental factors. The challenge for future forest managers and hydrologists is to design sustainable forestry systems for *Eucalyptus* species which minimize some of the adverse hydrological impacts that have been identified, whilst being compatible with the social and economic needs of the local people. All this needs to be set within the context of integrated water resource management.

An outline of the research findings from one study directed at resolving the eucalyptus concern is given below.

Land Use and Water Resource Field Study in Karnataka

To answer the questions raised about the environmental impacts of fast-growing tree plantations, and to devise ways in which some of the adverse effects could be minimized, studies of the hydrology of eucalyptus plantations, indigenous forest and an annual agricultural crop were initiated at four main sites in Karnataka.

Three of these sites were in the low rainfall zone (800 mm per annum) at the Devabal and Puradal experimental plantations near Shimoga and the Hosakote experimental plantations near Bangalore. The soils were of different depths, approximately three metres at Devabal and Puradal and greater than eight metres at the Hosakote site. The fourth site was at Behalli, in the high rainfall zone (2000 mm per annum) on deep soils (greater than eight metres) (Figure 5.4).

The research was carried out jointly by Indian and British organizations, the Karnataka Forest Department and Mysore Paper Mills and the University of Agricultural Science, Bangalore in India, and the Institute of Hydrology and the Oxford Forestry Institute in the UK.

Measurements were made of the meteorology, the plant physiology, soil water status, rainfall interception and the direct water uptake of individual trees using tracing measurements. Measurements were also made of the growth rates of the trees.

Water Use Results

Although the different experimental methods for determining water use operated over different time and space scales, the water use measurements obtained from the physiological, soil moisture, interception and tracing

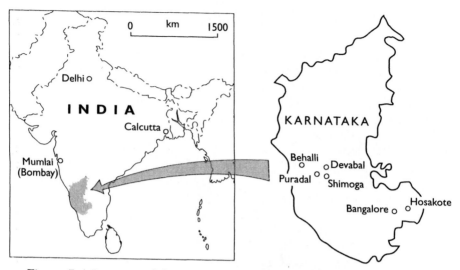

Figure 5.4 *Location of the eucalyptus environmental impacts study sites in Karnataka, southern India*

studies were generally much the same at all the sites. The deuterium tracing method, developed during the Karnataka project, proved to be a very effective and powerful method for determining transpiration rates of whole trees. This new method revealed a surprisingly 'tight' and simple relationship between the individual tree transpiration rate for young eucalyptus trees and the cross-sectional areas of their trunks (Figure 5.5). The relationship implies that, irrespective of the size of tree, sap velocities are the same. As water in the soil becomes scarce and water stress develops, the same linear relationship between transpiration rate and basal cross-sectional area appears to hold, even though the sap velocity decreases. This finding, together with the empirical relationship that in water-limited conditions volume growth is related to the volume of water transpired, provided the basis for a simple water use and growth (WAG) model (Calder, 1992b) that is applicable in India and other water-limited parts of the semi-arid tropics. Conventional methods for estimating the water use of vegetation are based on the assumption that evaporation is primarily determined by meteorological 'demand', a function of solar radiation and atmospheric temperature, humidity and turbulence. Conventionally the 'supply' side of the equation, where regulation is imposed through stomatal controls exerted by the plants, is considered to be of secondary importance. In temperate conditions where generally neither atmospheric demand is very great nor soil water severely limiting, these methods are very appropriate and successful. In the dry zones of India, on the other hand, it appears that the roles are reversed: water supply rather than meteorological demand primarily determines evaporation. The restrictions imposed by soil water availability, and for the young eucalyptus plantations, tree size, are so profound that, on a day-to-day basis, they limit the amount of evaporation that can take place. During the Indian dry season the meteorological demand imposed by the hot dry climate is far in excess of the water supply. Although the meteorological demand provides the energy and driving force for evaporation, it does not, when viewed over time scales of a few days, control the evaporation rate. This implies that for these supply-controlled conditions, evaporation rates can be estimated solely from consideration of the supply side of the equation without detailed knowledge of the demand.

This approach is of course very attractive from an operational viewpoint. By making use of these relationships and the working (and generally very good) hypothesis that volume growth is linearly related to the volume of water transpired, a water use and growth model has been developed whose only meteorological input is the daily rainfall. The WAG model provides the framework for the estimation of the water use from eucalyptus plantations in relation to age, spacing and growth rate. It also provides the framework for investigating the mechanisms that control the growth rates of the plantations in relation to water use – what is termed the water use efficiency. Through assignment of values, based on marginal costs, of the water consumed and the value per unit volume of the timber produced, it is possible to assess the economic returns of the plantation set against the real costs of the water consumed. These water-use calculations need also to build in interception losses and any other losses from the understorey or bare soil beneath the tree canopy. One feature of eucalyptus plantations is that although water consumption is high, transpirational water use efficiency is also high, so that for a given amount of water transpired the volume of timber produced is probably as high as that from any tree species. Bearing in mind that interception losses and understorey losses are essentially a 'fixed overhead' loss per year, not usefully employed, the high growth rates of *Eucalyptus* species means that these overheads are a small proportion of the total water consumed. On a plot basis the water use efficiency is probably as high if not higher than that of any known tree species.

The main water use findings were:

1   In the dry zone, the water use of young eucalyptus plantation on medium depth soil (three metres) was no greater than that of the indigenous dry deciduous forest.

2   At these sites, the annual water use of eucalyptus and indigenous forest was equal to the annual rainfall (within the experimental measurement uncertainty of about ten per cent).

3   At all sites, the annual water use of forest was higher than that of annual agricultural crops (about twice that of finger millet).

4   At the dry zone, deep soil (greater than eight metres) site the water use, over the three (dry) years of measurement, was greater than the rainfall. Model estimates of evaporation were 3400 mm as compared with 2100 mm rainfall for the three-year period. These results were later confirmed by an experiment carried out on an adjacent 'farmer's field' where measured soil moisture depletion patterns under eucalyptus, from the date of planting, were shown to be much greater than those under other tree species (Figure 5.5). They indicated that roots were penetrating the

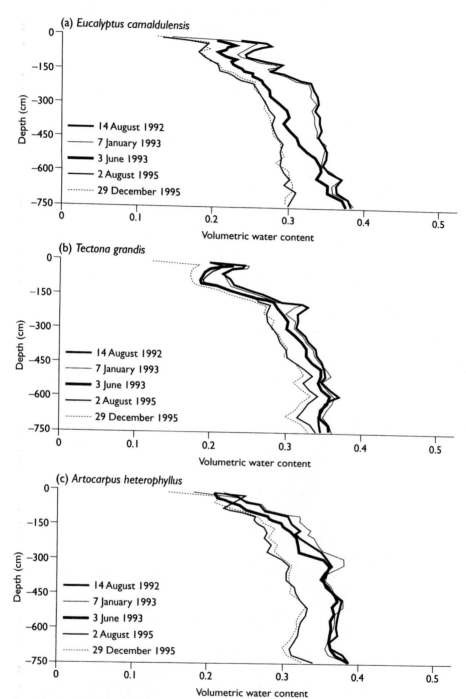

Note: Planting date: August 1992; profiles to 7.5 m depth

Source: Calder et al (1997a)

Figure 5.5 *Volumetric water content profiles recorded beneath the* E. camald-ulensis, T. grandis, *and* Artocarpus heterophyllus *plots, Hosakote, India*

soil at a rate exceeding 2.5 metres per year and were able to extract and evaporate an extra 400 to 450 mm of water in addition to the annual input of rainfall.

5   At none of the sites was there any evidence of root abstraction from the water table. Where this has happened in Australia at a site where the water table is relatively shallow, there are reports of annual eucalyptus water use of 3600 mm in areas where the rainfall is only 800 mm (Greenwood et al, 1985).

Erosion Results

Contrary to previous expectations and earlier published results, the Karnataka study established that some tree species are much worse than others in their potential for inducing raindrop splash-induced soil erosion.

Measurements of the modification of raindrop size by different tree canopies were obtained using a purpose-built opto-electronic device known as an optical disdrometer. Its use in India immediately provided new and hitherto unforeseen results on drop-size modification and erosion potential that have great relevance to the choice of tree species for erosion-sensitive environments.

Experiments at the Puradal and Hosakote sites show that not only are there large differences in the drop-size spectra beneath different tree species, but that each species, irrespective of the drop size of the incident rain, has a spectrum characteristic of that species (see Figure 1.5).

For the three different trees studied, Caribbean pine (*P. caribaea*), eucalyptus (*E. camaldulensis*) and teak (*T. grandis*), the median volume drop diameters of the throughfall ranged from 2.3 to 4.2 mm. The corresponding drop kinetic energies beneath the different trees, assuming that the drops reach terminal velocity, differ by a factor of about nine between Caribbean pine with the least, and teak with the greatest, kinetic energies.

The erosivity of natural rainfall is related to its intensity. As rainfall intensity increases, the size of raindrops generally increases also. There is a defined relationship between drop size and rainfall intensity from which the erosivity of the characteristic spectra may be related to the erosivity of rainfall of different intensities. The characteristic spectra for Caribbean pine, eucalyptus and teak correspond to the spectra of natural rainfall that would be expected with intensities of 48 mm per hour, 200 mm per hour and an essentially infinite rainfall intensity, respectively. These intensities can be interpreted as the 'break even' intensities at which the erosive power of rainfall of higher intensity would be moderated by the canopy.

Clearly, the canopies of Caribbean pine and eucalyptus, species exotic to India, will have a moderating effect for high-intensity storms. But there is no naturally occurring rainfall intensity for which the indigenous teak can exert a moderating effect; it will always produce a modified spectrum for drops falling at terminal velocity that is more erosive than the natural rainfall.

The canopy of the tree is not the only canopy affecting drop-size modification and erosion. The presence of an understorey canopy, close to the ground where drops cannot reach terminal velocities, is very effective in ameliorating splash-induced erosion. Where the understorey has been removed through fire or biological competition, the potential for erosion is very much increased (see Plate 1.1).

Water Use Efficiency

The recognition that water is a valuable resource in its own right and that forests generally evaporate more water than other crops, provides a powerful incentive for trying to improve the water use efficiency of plantation forests.

Both growth rates and water use efficiency of the eucalyptus plantations in the dry zone in India are low by world standards. To some extent climatic factors, which are not amenable to manipulation, are responsible for the low water use efficiencies. (High vapour pressure deficits, through increasing atmospheric demand, are known to decrease water use efficiency, while high temperatures, leading to high rates of maintenance respiration, also have a depressing effect on water use efficiency.) Nevertheless, it is believed that there is still great potential for improvements through, for example, species selection, removing nutrient and water stress, and improved silvicultural practices such as optimal spacing and weeding.

These aspects were studied in a Controlled Environment Facility at the Hosakote site where three tree species – *E. camaldulensis, T. grandis* and an indigenous species *Dalbergia sissoo* – were grown on 10 by 10 metre plots under different conditions of water and nutrient stress. The different stresses were imposed by water treatments of 0, 2.5, 5.0, and 7.5 mm applied each day, irrespective of rainfall, through a drip irrigation system and nutrient treatments of nil, one and four times a standard application rate which were applied in the dry form on all plots twice a year.

Results from the first two years of the experiment showed up to five-fold increases in growth rates on plots which received both water and fertilizer treatments compared with the control plots (Figure 5.6). Although water applications alone lead to large increases in growth rate, there was no evidence to show that purely transpirational water use efficiency was improved.

Water Resource Implications

The hydrological studies carried out in southern India on plantations of exotic tree species, indigenous forest, and an agricultural crop show a varied and complex pattern of hydrological impacts. In summary, the water resource implications of afforestation with exotic species are as follows:

1   Erosion. The net rainfall size spectra associated with such exotic species as *P. caribaea* or *Eucalyptus camaldulensis* make their plantation preferable, from a soil conservation perspective, to *T. grandis* which has a characteristic net rainfall spectrum of potentially much greater erosivity. The common occurrence of fires beneath teak plantations, which destroys the protective understorey, is another reason for not planting teak on sites sensitive to erosion.

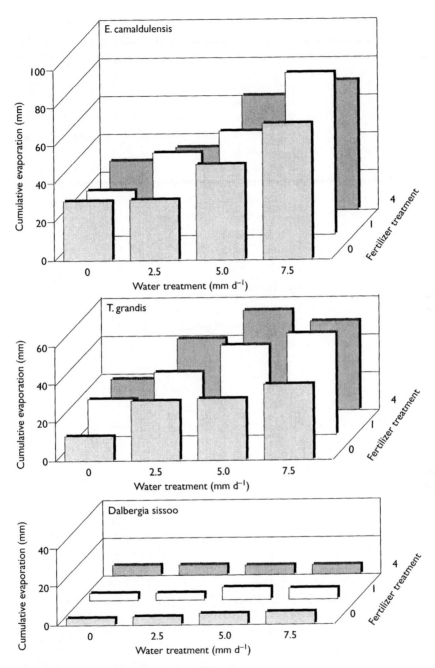

Note: The trees were planted in October 1991.

Figure 5.6 *Stand volumes recorded in September 1993 for* E. camaldulensis, T. grandis *and* Dalbergia sissoo *(top to bottom) for a range of water and fertilizer treatments*

2   Water use. At the Devabal and Puradal sites where the water use of eucalyptus plantations has been compared with that from indigenous forest, there is no evidence that *Eucalyptus* species use more water than the indigenous dry deciduous forest. They do nevertheless use more water than a typical annual crop – about twice as much as ragi, a finger millet.

3   Water use. The Hosakote findings, which show much higher water use from young eucalyptus plantations than from young plantations of other tree species, and water use greater than the rainfall input, have important and serious water resources implications.

4   Water use efficiency. Although the water use of eucalyptus plantations is much higher than those from other tree species the water use efficiency, expressed on a plot basis, is also much greater from the eucalyptus plantations. For the same amount of water consumed, on a plot basis, a higher return in terms of useful biomass will be achieved from the eucalyptus plantations.

Sustainable Management Systems

From the research experience outlined earlier, it is suggested that the potentially adverse aspects of plantation forest practices can be curtailed through the adoption of the following:

1   Rotation. Where soil water 'mining' occurs then one strategy which may prove advantageous, particularly on deep soils, would be to rotate *Eucalyptus* with agricultural crops. A five-year period under an agricultural crop should allow the soil water reserves, depleted by say ten years of forestry, to be replenished. From studies in other arid zones of the world, there is evidence that deep-rooted trees bring up nutrients from deep soil layers to the surface. If this is true of *Eucalyptus* species in India, then there would be dual benefits from rotation; the trees would replace nutrients the agricultural crops remove whilst the agricultural crops would replace water that the trees have removed.

2   Patchwork Forestry. The forest water-use results indicate that recharge to the groundwater under large areas of either plantation or indigenous forest in the dry zone in India is likely to be small and will not, on average, be more than ten per cent of the rainfall. However if plantation forests were grown as a 'patchwork', interspersed with annual agricultural crops, much of the adverse effects on the water table would be alleviated as up to half the annual rainfall should be available for recharge under the agricultural crops.

3   Irrigation. In theory it would be possible to optimize a 'patchwork' design with irrigated areas of forestry. It should be possible to grow the same volume of timber, using irrigation, on one-fifth to one-tenth of the usual land area leaving the rest for rainfed agriculture. There may also be economic advantages of this type of scheme. If the plantations were located close to the pulp mills, transport costs would be minimized which could halve total production costs.

AUSTRALIA AND SALINIZATION

Background

> *'Salinity and its avoidance or management has been an enigma in Australia. On one hand there has been a wide understanding of the inherently salty nature of the environment and the inevitability of salinity. On the other, there has been a high propensity to ignore the problem and believe that for some reason salinity will not develop in any specific area.'*

This quotation (Robertson, 1996) is a vivid illustration of the frequently occurring situation worldwide where, when land use and water-resources decisions need to be taken promptly, and although there is no lack of research or knowledge, the issues are fudged and deferred. No rational reasons are given although self interest and short-termism may be suspected.

Robertson makes the case that the early explorers in Western Australia regularly encountered and reported saline land and water: *'with exception of rivers flowing out from the Darling Ranges, WA rivers are beds of salt, pools of brine and brackish water'*. The relationship between land clearing and salinity was widely recognized by the turn of the century – where natural forests on catchments had been ringbarked, increased salinity had been recorded in streams and reservoirs. In an early publication, Wood (1924) clearly demonstrated that there was both salt stored within the landscape and in groundwaters and that when native vegetation was cleared, this resulted in increased salts in the upper soil horizons and surface water.

Robertson concluded *'Unfortunately, despite this and a more of knowledge, research and experience, the "not here" syndrome has prevailed and we are still seeing salinity expand'*. Some authorities ominously believe that if steps are not taken quickly, as much as 48 per cent of Australia's agricultural land will go out of production, irreversibly, within the next few decades.

The Salinization Process

It is now generally believed that salinization of the landscape has occurred over geological timescales. Windblown salts from the ocean and salt in rainfall have led, in the generally low rainfall and high evaporation climate of much of Australia, to a gradual build up of salt in the soil profile. Hingston and Gailitis (1976) showed that accretion of salt from oceanic aerosols was between 100–200 $kg.ha^{-1}.yr^{-1}$ in high rainfall coastal areas, falling to 10–20 $kg.ha^{-1}.yr^{-1}$ at a distance of 300 km from the coast. At these rates of accretion 10,000 years would have been sufficient to account for the measured salt concentrations in the coastal zone, but continuous accretion from the late Pleistocene would have been necessary to account for the measured concentrations in the more arid interior. The indigenous *Eucalyptus* dominant vegetation, widespread over much of Australia, has established an evaporative regime that is almost in exact balance with the rainfall input. The balance is such that salts within the root zone are flushed through to the saline water-table below, but the evaporation rate from the natural forest is sufficient to prevent the water-tables rising, either into the root zone and killing the vegetation, or rising to the surface and seeping into watercourses. In this precariously balanced ecosystem, man's intervention has had serious consequences. Clearance of the indigenous forest has realized the danger of raised water-tables leading to incidents of saline seeps developing in low-lying areas where the water-tables reach the surface. The huge civil engineering exercises of the 1950s and 1960s, sometimes, as in the Snowy Mountains Project, involving the reversal of flow direction of the rivers, have resulted in mixed blessings. Although in the short term agricultural productivity has been increased, this has been at the expense of longer-term environmental and ecological damage. Excess irrigation waters from these schemes are a major contributing factor to higher water-tables, increased leaching of salts into the watercourses and the possible irreversible damage that is being done to the ecology and environmental health of the major river systems such as the Murray-Darling.

Management Options

A range of options has been recognized as beneficial in controlling and, hopefully, reducing dryland and river salinity. These options can loosely be grouped under three headings: land management, engineering solutions and economic tools.

Land Management

Replanting trees and shrubs in the recharge areas of a catchment will increase the evaporation, through increased interception and probably also increased transpiration, and reduce recharge and groundwater levels. This will alleviate salinity problems in seepage areas by both lowering water-tables and lowering the level at which seepage takes place, hence reducing the land area affected, and also by reducing the volume of seepage waters. The Western Australian Department of Agriculture (1988) recognizes that while it may not be economic to carry out large-scale planting, it recommends planting on identified specific recharge areas within a catchment, especially if these areas are small and do not produce economic crops. In and around the highly saline seepage areas, replanting with salt-tolerant shrubs such as saltbush (*Atriplex*) for forage production has been advocated (Malcolm, 1990) and is being incorporated into whole farming systems.

In areas where dryland salinity problems are less extreme, changes in cropping pattern have been advocated. The replacement of shallow-rooted grasses with deep-rooted alfalfa was found to be very effective for salinity control in the Northern Great Plains of the USA (Halvorson and Reule, 1980). In Australia the growing of deep-rooted lupins in the rotation of grain crops has been advocated rather than short-rooted clover, for the same reasons. Perennial pasture has also been shown to be effective in reducing recharge. A rotationally grazed stand of lucerne used 433 mm of water annually as compared with 231 mm for an adjacent wheat crop (Western Australia Department of Agriculture, 1988).

Engineering Solutions

The state and federal governments of Australia have traditionally favoured engineering solutions to control salinity problems. Various engineering schemes have been tried or proposed. These include the interception of saline groundwaters and the diversion of the excess water (returns) from irrigation schemes to evaporation pans or directly to the major rivers, such as the Murray River. Direct pumping of groundwater to lower water-tables has also been tried with some success in some agricultural areas, but a major problem with this type of solution is the disposal of the saline effluents. Disposal to rivers is the common option, but often serves only to salinize the vital water supply for the less fortunate users who happen to be located downstream. If downstream users also drain their fields or pump groundwater in a similar fashion, rivers will undergo progressive salinization until their ecology is destroyed and lower reaches become unfit for human use or irrigation. Piping the effluents for discharge into the sea has been considered, but is generally thought not to be cost effective. The combination of engineering solutions with land use management has been claimed to be most effective. Some success in salinity control in the wheatbelt of Australia has been obtained through the combined use of land drains for groundwater interception together with eucalyptus plantations in recharge areas.

Economic Tools

In most countries irrigation water has an artificially low price (see Chapter 4) and is heavily subsidized by governments. Australia is no exception. In the early 1990s subsidisation was of the order of A$300 million a year (Simmons et al, 1991). Increasingly it is recognized that subsidized and low prices for water lead to inefficiencies in use that contribute to waterlogging and salinization problems. Increasing the price for irrigation water is one method that can be adopted for encouraging efficient water use. In Victoria, gravity surface-water supplies were being charged at A$12.0 per 1000 m^3 in the early 1990s, but the policy is now to increase charges at two per cent above inflation until the full supply cost of A$22.5 (at 1989 values) is reached (Evans and Nolan, 1989).

Another method, which has been used in the USA and also in some states of Australia, is the Transferable Water Entitlement (TWE) or transferable water right. This is a mechanism by which a market for water can be achieved by allowing entitlements to be bought and sold without the necessity of buying and selling the accompanying land. The use of the mechanism is expected to increase efficiencies in a number of ways including the transfer of water to higher value uses and higher valued crops. It is also expected to lead to the increased adoption of water-saving irrigation technologies, because the saved water can then be sold. Decreased use of irrigation on land which is poorly suited and where economic returns are low, perhaps because of existing waterlogged or salinized conditions, might also be anticipated.

Another economic tool involves salinity credits. The theory is that Murray-Darling basin states can earn salinity credits by carrying out land management and engineering schemes that reduce salinity in the Murray-Darling River. They can then construct drainage and aquifer pumping schemes which increase salinity in the river network, provided that the salinity increase does not exceed their credit. The flaw in this approach, as pointed out by Macumber (1990), is that the amount of salt that needs to be disposed of from the aquifers is far in excess of any possible credits that could be achieved.

Research and Integrated Management

Australia has committed much in the way of research funds to investigate and address the issue of salinization. The equity, economic, ecological and sustainability issues related to salinization are all researched and fairly well understood. However, although the technology and awareness are in place, the salinization problem has not even been contained let alone resolved.

Why this should be so in one of the most environmentally and ecologically aware countries in the world is not immediately obvious, and does not auger well for the blue revolution. Where does the problem lie? Could it be that the direct incentives for the different stakeholders are not in place? Perhaps the farmer, attending a meeting on integrated catchment management, who enthuses about the benefits of land management as a means of salinization control, may ask himself on the way home *'what is in it for me?'*. Why should he forgo production on his most productive land by planting trees when the benefits may well accrue not to him, but the landowners downslope? Or perhaps the farmer thinks that by delaying another year will not make much difference and he can get another year's grain crop. Perhaps the politicians feel that there are no votes in pressing through measures which will cost their electorate hard cash in the short term – even though the long term benefits are patently obvious to all.

This is perhaps the greatest challenge of the blue revolution: to develop methods to recognize and capture the human dimension and the aspirations and motivations of the different stakeholders. From an appreciation of the different viewpoints it may be possible to put in place management philosophies which can provide recognizable incentives to all parties. The Comprehensive Assessment of the Freshwater Resources of the World, produced for the Commission for Sustainable Development (UN,1997), hails the Murray-Darling Basin Commission as a successful example of the integrated water management approach:

'The Murray-Darling Basin covers one-seventh of Australia, and accounts for half the country's gross agricultural production. As demands for water increased, reservoirs were constructed to increase the available supply to individual states. In recent years, use approached the sustainable yield of the basin as a whole, and pressure mounted for sharing the resource between jurisdictions. In 1985, a Basin Commission was formed and in 1989, agreement was reached on sharing. The next issue requiring resolution was soil salinity that had the potential to expand to 95 per cent of the total irrigated area within 50 years. The three upstream states were the primary beneficiaries of water diversion, while the damage caused by salinity was most severe in the downstream state. An agreement was reached on joint funding of remedial measures and collaboration was initiated, driven primarily from the community level. Action has been under way for four years, and the spirit of collaboration continues as a demonstration of integrated water management success.'

This view is heartening and it is certainly to be hoped that the Commission is successful in its task.

Unless this can be achieved and the issues are really treated and resolved in an integrated way, the ominous conclusions of Robertson, foretelling of an environmental disaster of enormous proportions, will almost certainly happen. There is clearly little time to waste if an irreversible disaster is to be prevented. Already some experts are proclaiming that the Murray-Darling basin is irretrievably lost as a freshwater system and that it should now be regarded as a saline conduit to remove irrigation drainage waters and saline seeps to the ocean.

USA: PIONEERING APPROACHES AND PRACTICE

The zeal of the pioneer is evident in North America's history of river and water resource management. The influence and dynamism of the engineering community in 'taming the river' and 'draining the swamp' was such that there are now more than 50,000 major engineering structures – dams of over 25 feet in height – within the USA and many of the major river systems have reaches which have been 'straightened' or canalized. The dams served America well in terms of flood control, water supply and hydroelectric generation. But by the late 1960s and early 1970s a new, more questioning, attitude towards the benefits of further engineering developments was beginning to influence decision making. This coincided with the beginnings of the environmental movement, activism by conservationists and landowners and a growing awareness of resource scarcity. The better appreciation of the environmental and social costs associated with many engineering developments has profoundly changed America's approach to these developments. Indeed, not only is it now more difficult to obtain the necessary agreements to allow new engineering developments, but in some cases river rehabilitation programmes have been initiated which are actively reversing the alterations that had been made to river and water resource systems, to pursue a more 'natural' environmental regime. Arguably, America, which was once at the vanguard of the engineering approach to water resource development, is now at the vanguard of the movement towards a more natural and more environmentally sustainable and socially acceptable water resource regime. The power of this new movement is evident through the new approaches demonstrated in the sustainable management of aquifer systems such as the Olgallala, the very strong environmental focus now displayed in the management of the Great Lakes, and the social and environmental orientation of both the Heritage Rivers Initiative and the Kissimmee River restoration programme, which are described below.

The American Heritage Rivers Initiative

In the State of the Union Address of 4 February 1997, President Clinton pledged:

> *'Tonight, I announce that this year I will designate 10 American Heritage Rivers, to help communities alongside them revitalize their waterfronts and clean up pollution.'*

Following up on this pledge, President Clinton signed an executive order establishing the American Heritage Rivers Initiative (see Appendix), a new programme to help communities restore and revitalize waters and waterfronts. The initiative has three objectives: natural resource and environmental protection, economic revitalization, and historic and cultural preservation. It encourages communities to come together around their rivers and develop strategies to preserve them. It was expected that Americans would look toward rivers as sources for improving community life and that the American Heritage Rivers Initiative would *'integrate the economic, environmental and historic preservation programs and services of federal agencies to benefit communities engaged in efforts to protect their rivers'*.

The new approach was further illustrated in a later speech made in North Carolina (President Clinton, 30 July 1997), to designate 14 American Heritage Rivers:

> *'Who are we, such brief visitors on this Earth, to spoil the rivers and other treasures that were here so long before we arrived? We must work together, as this community has done for generations, to preserve all these sacred gifts for all time.'*

The American Heritage Rivers programme is structured to support *'outstanding community-based efforts designed to ensure the vitality of the river in community life for future generations'*. It was set up as a locally-driven programme on the assumption that the stakeholders, the local communities, 'know best what they need'.

Once a heritage river is designated, a full time contact, called a 'River Navigator', coordinates efforts to match community needs with available resources from the existing programmes.

Kissimmee River Restoration

Before engineering works for flood control, the Kissimmee River meandered from its origin in Lake Kissimmee to the northern shore of Lake Okeechobee. The extensive 18,000 hectare floodplains of this Florida river provided a wide range of wetland habitats with over 35 species of fish, 16 species of wading birds, 16 species of waterfowl, river otters, and many species of invertebrates, amphibians, and reptiles.

The flood control project was conceived and designed between 1954 and 1960, a time which predated much of the environmental law that came about in the USA in the late 1960s and early 1970s; the Clean Water Act of 1972, the National Environmental Policy Act of 1969, and the Endangered Species Act of 1973.

At the time of authorization, and when construction began in 1961, environmental concerns were expressed, but governmental water resource management processes were, at that time, relatively single minded. The aim was simple: flood damage reduction in the most cost-efficient fashion.

That was what was done. As part of the Central and Southern Florida Flood Control Project, the US Army Corps of Engineers carried out the engineering works which improved the navigation and flood control by transforming the natural 166 km winding path of the Kissimmee River into a 90 km long, 9 m deep, 100 m wide canal, known today as the C-38 canal.

The project hugely altered the hydrology and ecology of the river basin. It was widely seen as an environmental disaster and efforts to restore the river were being contemplated even before the project finished. The engineering works caused between 12,000 and 14,000 ha of wetlands to drain and dry up, with the consequent loss of the wetland flora and fauna. The changes in the flow regime also created conditions suitable for invasive plant and animal species, many of which were regarded as pests.

The Act of 1976 initiated a series of state and federal initiatives and research programmes aimed at gaining the knowledge required to restore the integrity of the river in order to retrieve some of the lost environmental benefits. After 15 years of research and planning, the state of Florida finally adopted, in 1990, the South Florida Water Management District's restoration plan and sought authorization and joint funding for the plan from the federal government.

With enactment of the 1992 Water Resources Development Act, Congress approved the Corps of Engineers recommendation to undertake a river restoration programme, which was aimed, essentially, at getting the Corps to fill in the canal they had originally dug.

This, perhaps the world's first major watershed restoration project designed to reverse the impacts of earlier engineering works, is now under-

Plate 5.3 *Kissimmee River, Florida, USA*

way. Backfilling of about 15 km of the canal in Phase I of the restoration was scheduled to begin in March 1999.

With regard to the Kissimmee River Restoration, M. K. Loftin, Assistant Director, Water Resources Division and Project Manager for the Kissimmee River Restoration, South Florida Water Management District stated at the Kissimmee River Restoration Symposium, held in Orlando, October 1988, that *'environmental problems are emotional; environmental issues, political; and environmental solutions, technical'* (EPA, 1992). More projects like the Kissimmee Restoration may be required to achieve the desired balance of economic, environmental and social objectives in IWRM.

LAND USE, PRODUCTION, THE ENVIRONMENT AND AMENITY IN THE UNITED KINGDOM

Although the earliest studies of water and land use issues can perhaps be ascribed to Switzerland and France in the 19th century and Hubbard Brook in the USA in the early 20th century, the UK began to make a significant contribution by the 1950s. The catchment studies in East Africa (Pereira et al, 1962) funded by the British ODA, in the early 1950s, the catchment and process studies initiated by Frank Law in the Yorkshire Pennines, and the later inception of catchment and process studies at Plynlimon and Thetford, were all seminal in the development of land use/hydrological studies in the UK. The theoretical framework for measuring and estimating evaporation provided by Howard Penman and John Monteith, both Fellows of the Royal Society established British science in this area, a contribution arguably unequalled anywhere else in the world. British scientists have also made major contributions to research into land use hydrology elsewhere in the world. The UK has also been a world leader in the development of methodologies for IWRM and was perhaps the first country to fund the development of modelling methodologies which link the water resource, economic and ecological aspects of land management (see Chapter 6).

Yet although the funding of research may be a requirement for proper water resource and land use management, is it sufficient? If the Rural White Paper on England (1995) is to be used as a criterion, the answer must be 'no'. Although this well meaning White Paper captures the public desire for the amenity aspects of rural land use, there is a singular unawareness of the interlinking water resource issues. Is this the fault of the scientists for not disseminating their results or the policymakers for not taking the trouble to make themselves aware of them? Have the scientific results been poorly and ambiguously presented? Has there been a lack of involvement of stakeholders in defining and carrying out the research? Has the popularist scientific press been slow to take up the issues? The answer is probably "yes' to all. Although all the ingredients have been present, the proper mix and end-product have not been obtained. The poor linkage between land-use research in the uplands and its application in land-use management is decried by Newson (1997) in his seminal book on land and water issues: 'Land, Water and Development'. He states that

> '*three years after the acceptance of a paper on the hydrological impacts of afforestation in the UK by the hydrological community, the Secretary of State for the Environment in the UK Government made the following statement in Parliament: "As regards afforestation, its percentage and its effects on catchment areas...I am advised there is a lack of clear scientific evidence."*'
> (Hansard, 21 March 1980)

Newson has attempted to '*put the knowledge base of hydrology in a policy context*' through later publications (Newson, 1990, 1991, 1992b) and these have resulted in a slow, but measurable, policy readjustment to the original research. This book continues that theme in the hope that the UK, together with countries such as South Africa, can take benefit from the results of its own land-use research.

An outline of some of the contributions to land use hydrological research, tracing the conflicts that have arisen in the uplands and lowlands of the UK and the development of modelling methodologies to assess the impacts of the land use changes, are given below. The development of modelling methodologies for addressing forest-impact issues is given at some length as these methods have subsequently been found to have wide application and have formed the basis for land-use impact modelling in other temperate and tropical regions of the world.

Impacts of Water Resource Developments on the Environment

In recent years some of the most controversial water-related conflicts in the UK have arisen not so much from the effects of environmental and land use changes on water resources, but from the effects of water resource developments on the environment and ecology.

Of particular concern have been the groundwater and river abstractions by water companies (see Chapter 4) which have led to low summer flows in rivers and, in some circumstances, dried-up river beds on what were normally perennial rivers. Increasingly the water resource planner must really take into account and balance the needs of the environment with the economics of water resource developments and water demands (Smith, 1997). But this balance may not be easy to achieve. Foster and Grey (1997) state that sustainable groundwater management requires maximizing the use of aquifer storage to reduce water-supply costs while limiting environmental impacts, and maximizing groundwater protection to reduce water supply treatments while not unduly restricting land-use activities. They argue that achieving this balance is difficult because groundwater systems are complex and sometimes slow to react to change.

Environmental bodies have been active in developing initiatives and programmes which are aimed at reducing the environmental impact of these actions. English Nature's wildlife and freshwater initiative details a plan of action for conserving Sites of Special Scientific Interest (SSSI) whilst the

Ministry of Agriculture Fisheries and Food has a water level management-plan programme in force which is specifically directed at protecting wetland SSSIs. The Royal Society for the Protection of Birds (RSPB) has published a report entitled *'Practical Implications of Introducing Tradable Permits for Water Abstractions'* which claims that the present method of regulating abstractions, using abstraction licensing, does little to discourage the wasteful use of water. Tradable Permits for Water Abstractions would operate in a similar way to the Transferable Water Entitlements or transferable water rights that are used in America and some states of Australia (see above), and would be expected to increase the efficiency and allocation of water use by farmers where presently some have too much and others not enough for their requirements. The RSPB believes that tradable permits would reduce the pressures on the environment and help maintain water levels at key wetland sites. Studies have been made to examine the practical extent to which water rights can be traded within the existing regulatory system, which have led to proposals being made to deregulate existing legislation to allow trading to proceed (Streeter, 1997).

Increasing pressure from the public and environmental bodies will ensure that land and water resource planners will have to take full account of nature conservation and the environment in all future planning decisions. The environment secretary has recently accepted recommendations from the Joint Nature Conservation Committee for the addition of 33 plant and animal species to the list of those receiving protection under the Wildlife and Countryside Act. This now includes water voles whose range and numbers have declined rapidly because of the introduction of American mink. Developments which disturb the riverbank-homes of water voles will now be illegal. But achieving the balance between water resource developments, and the impacts of these developments on basin economics, ecology and the environment is clearly no easy task and tools are being developed (see Chapter 6) which can assist land use and water resource planners in this task.

Upland Afforestation Conflicts

Since the mid-1950s, when Frank Law (Law, 1956), engineer to the Fylde Water Board, published the results of his studies from Stocks Reservoir in the Lancashire Pennines, a controversy has continued over the effects of upland afforestation on evaporation and water resources of the UK. Originally the concern centred on the effects of spruce afforestation of upland moorland water catchments that were being used for supply purposes; then, by the late 1970s, the deleterious effects of afforestation on hydroelectric power generation were recognized. By the late 1980s the hydrological effects of extensive larch plantations became an important issue, resulting from food surpluses being generated within the European Community, the reductions in subsidies to farmers and the consequent probability of marginal agricultural lands (not always in the uplands) being afforested.

By the late 1970s, studies carried out by the Institute of Hydrology and other organizations had validated Law's conclusion of enhanced evaporation rates from upland forests. Nevertheless doubts were expressed that because most of the experimental studies had been carried out in the relatively warm climates of Wales and England, primarily on grass and forest vegetation, the results could not necessarily be applied in Scotland. In Scotland the climate is cooler and snow can form a significant component of the annual precipitation, and heather rather than grass moorland predominates. Further studies, involving both process and catchment experiments were executed to quantify these effects through process studies at a number of upland sites and catchment experiments located at Balquhidder in central Scotland (Calder, 1990). Although quantity concerns were originally predominant, questions of water quality later came to the fore. In the high-pollution climate of the UK, the process responsible for the high evaporative losses from forests – enhanced aerodynamic transport arising because of the high aerodynamic roughness of forest canopies – is also responsible for the much higher rates of gaseous and particulate pollutant deposition to forests. These high deposition rates meant that forested catchments in the UK, and the streams emanating from them, were generally more acid, than those from upland grassland catchments.

Upland Land Use Research at Plynlimon

An example of the use of process studies in conjunction with catchment experiments to examine the effect on water use of land use changes associated with forestry, is provided by the studies carried out by the Institute of Hydrology at Plynlimon, Central Wales. The catchments are located at the source of the Rivers Wye and Severn in steep upland topography (Plate 5.4).

The annual precipitation is of the order of 2400 mm (Table 5.2), and is distributed fairly evenly throughout the year. Most of the precipitation occurs as rain; the snow contribution is very variable, but averages 5 per cent per year. The rain is mostly of low intensity, generated from frontal systems enhanced by the orographic effect of the hills. Streamflow is perennial although storm runoff forms a major proportion of the flow. The soils are predominantly peaty, overlying mudstone and shale drifts on the slopes. Peat of up to three metres in depth occurs on the hilltops and in the valley bottoms where it overlies glacially deposited boulder clay. The Wye catchment is under grass cover; 70 per cent of the Severn catchment is under coniferous forests, mostly Norway and Sitka spruce, with 30 per cent under moorland grass.

Measurements at Plynlimon

At Plynlimon the processes controlling transpiration have been measured using a number of techniques. These include the use of a 'natural' lysimeter (see Plate 3.1), together with neutron probe and tensiometric measurements of soil moisture (Calder, 1990), plant physiological measurements of stomatal

Plate 5.4 *The forested Severn catchment at Plynlimon, Central Wales*

conductance and leaf water potential, and 'tree cutting' measurements of water uptake from excised trees (Roberts, 1978).

Table 5.2 *Measurements from the Plynlimon forest lysimeter, February 1974 to October 1976*

Period	Precipitation (mm)	Interception (mm)	Transpiration (mm)
6 Feb–31 Dec 1974	2328	685	289
1 Jan–31 Dec 1975	2013	529	335
1 Jan–1 Oct 1976	1103	366	277
Total	5444	1580	901

Note: The precipitation was recorded at the nearby Tanllwyth gauge within the Severn catchment.

Source: Calder et al (1982).

A number of techniques were also used to measure the evaporation arising from the proportion of the rainfall that is intercepted by the vegetation and re-evaporated before it reaches the ground, ie, the interception. These techniques included conventional throughfall troughs and stemflow gauges, plastic-sheet net-rainfall gauges (Calder and Rosier, 1976) and gamma transmission measurements (Calder, 1990).

Plynlimon Process Study Results

These process studies conclusively demonstrated that at Plynlimon the reduced runoff per unit-area from the forested catchment is principally the result of the increased interception losses from the forest. The higher interception losses are generated because of the increased turbulence and lower aerodynamic resistance to the transport of water vapour and heat between the forest surface and the atmosphere. This leads to higher evaporation rates from the forest in wet conditions compared with grassland. The enhanced evaporation rates occur both during rainfall and immediately afterwards from the wetted vegetation surface; about half the interception loss occurs during rainfall.

The transpiration from the forest is typically about ten per cent less than that from grassland as a result of physiological controls imposed by the forest (lower stomatal conductance). Because of the high and seasonally even rainfall climate at Plynlimon, periods with soil moisture deficits sufficient to limit transpiration are not common.

Upland Research and Land Use Evaporation Models

The two results – that interception losses from tall vegetation are likely to be higher than those from short vegetation and that, when soil moisture is non-limiting, forest transpiration is likely to be similar to but less than that from grass – have general significance. They can, with few qualifications, explain the results from the majority of the world's 'forest/grass' catchment experiments (see Hewlett and Hibbert, 1967). A third generalization, applicable in more arid regions, where large soil moisture deficits occur, is that the greater rooting depth of forests, and greater soil water availability to forests, compared with grass and agricultural crops, leads to higher transpiration rates from forests. So during drought periods, evaporation losses from forests may also be higher than those from grasslands, but for different reasons.

The qualifications to these generalizations mainly concern the observations from four south-east Australian catchments (Langford, 1976) which, after forest fires, showed a decrease in runoff in subsequent years of 24 per cent as compared with runoff from a catchment that escaped the fire. This apparently anomalous result, one of the few examples of forest removal decreasing runoff, was explained by Greenwood (1992) in terms of the unique forest structure. He pointed out that the forest, which was composed predominantly of exceptionally tall (98 m) *Eucalyptus regnans* (some of the world's tallest trees), had a minimal canopy which reduced both interception and transpiration losses. Following the fire, the germination of seeds and the subsequent regrowth rapidly led to the leaf area per unit ground area exceeding that of the former forest canopy.

To make use of the upland land use research findings for water resource management required the development of evaporation models. The development of these models, which now have a much wider application outside the UK uplands, is described below.

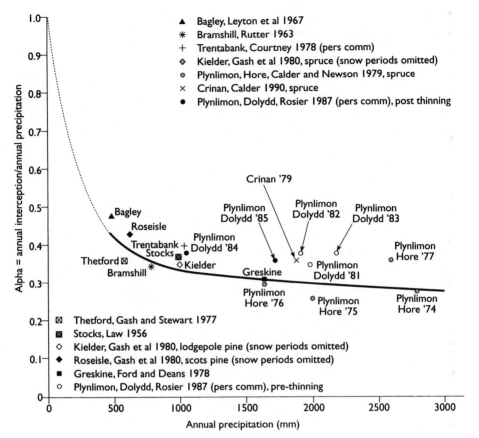

Source: Calder (1990)

Figure 5.7 *Observations of the annual fractional interception loss from forests in the UK*

The 1979 Forest Impact Model

Based on the upland research findings, a semi-empirical model was developed by Calder and Newson (1979) for estimating the annual and seasonal differences in runoff from afforested, upland catchments in the UK, which were previously under grass cover. These catchments receive rainfall throughout the year and periods with large soil moisture deficits are uncommon. This original forest-impacts model was parsimonious in both data requirements and in the number of model parameters and was designed to be of practical value as an operational tool for water resource assessment. This ethos has been preserved in later developments of the model which now have a wider range of applications. The original (1979) model requires information on annual or daily rainfall, annual or daily Penman (E_T) estimates of evaporation, and the proportion of the catchment with complete canopy coverage.

For the calculation of annual evaporation, the assumptions inherent in the method are that:

1   Evaporation losses from grassland are equal to the annual Penman potential transpiration estimate for grass, E_{Ta}.
2   Transpiration losses from forest are equal to the annual E_{Ta} value multiplied by the fraction of the year that the canopy is dry.
3   The annual interception loss from forest, with complete canopy coverage, is a simple function of the annual rainfall, P_a (Figure 5.7).
4   Soil moisture deficits are insufficient to limit transpiration from grass or trees in this (wet) area of the UK.

This leads to the equation for the calculation of annual evaporation:

$$E_a = E_{Ta} + f(P_a\alpha - w_a E_{Ta})$$

where:

α = the interception fraction (35–40 per cent for regions of the UK where annual rainfall exceeds 1000 mm),
w_a = the fraction of the year when the canopy is wet ($\sim 0.000122 P_a$),
f = the fraction of the catchment area under forest cover.

Use of aerial photographs for upland UK forests has shown that, typically, for areas marked on maps as extensive forests, the 'f' value is about 0.66; the remaining area comprises roads, gaps between forest blocks, riverbanks, clearings and immature plantations with unclosed canopies.

The Calder-Newson model indicates that in the wet upland regions of the UK, annual evaporation rates from forested catchments (with 75 per cent of their area afforested, equivalent to 50 per cent canopy coverage) may exceed those from grassland by 100 per cent and runoff will be reduced, typically by about 15–20 per cent.

This simple model has been used to investigate the effects on water supplies of afforesting the catchments of the major UK reservoirs. It was also used, in the early 1980s, to provide information for the Centre for Agricultural Strategy's investigations into the feasibility of proposals to increase greatly the proportion of upland forestry in Britain. It has been used subsequently in many studies into the effects of afforestation on water resources in the UK.

Annual Model: Forest/Heather/Grass

Research on the evaporative characteristics of heather (*Calluna vulgaris*) in the UK uplands, has established that transpiration losses are less, but interception losses greater, than those from grassland. These observations (Table 5.3) suggest that the annual interception losses from heather can be estimated from an equation of the form:

$$E_a = \beta E_{Ta}(1 - w_a) + \alpha P_a$$

where $\beta = 0.5$ and $\alpha = 0.2$.

This equation indicates that evaporation from grassland and heather moorland will be similar when both are growing in regions that experience an annual rainfall of about 1250 mm. In regions with annual rainfall greater than this, the higher interception losses from heather will outweigh the reduced transpiration from the heather and the total annual evaporation from heather will be greater than that from grass; the converse is true for annual rainfall less than 1250 mm.

Daily Model: Forest/Heather/Grass

To investigate the seasonal variation of the effects of a land use change among forest, heather and grassland, the same approach was adopted with the incorporation of an interception model which operated on a daily timestep.

The model incorporated the two-parameter exponential relationship:

$$I = \gamma(1 - \exp(-\delta P))$$

where I is the daily interception loss (mm) and P is the daily precipitation (mm). With parameter values, $\gamma = 6.91$ and $\delta = 0.099$, the model was found to fit well with coniferous forest interception losses recorded at a number of upland sites in the UK.

Seasonal evaporative losses were obtained by summing the daily evaporation estimates, E_d, as given by:

$$E_d = \beta E_T(1 - w) + \gamma(1 - \exp(-\delta P))$$

Here the transpiration is estimated as the product of β, termed the transpiration fraction, a climatologically-derived daily Penman E_T estimate, and a term $(1-w)$ which represents the fraction of the day that the canopy is dry and is able to transpire (where w is the fraction of the day the canopy is wet = $0.045P$ for $P < 22$ mm; $w = 1$ for $P \geq 22$ mm). This equation has been validated by comparison with soil moisture and interception measurements at different sites in the uplands of the UK. Estimates of the α, β, γ and δ parameters for the different vegetation types and the sources from which they were derived are shown in Table 5.3.

Although the interception component of the model is based more on empiricism, rather than theoretical reasoning, the parameters can be attributed with some physical meaning. The γ parameter can be considered to represent the mean daily interception loss which would be obtained with an infinitely high value for the daily precipitation, essentially the maximum interception loss per day. The δ parameter governs the rate at which interception loss increases with increasing precipitation (whilst not allowing interception to exceed the precipitation). Consideration of more detailed experimental and physically based modelling studies of the interception process (Calder, 1996b, Calder et al, 1996), which led to the development of the Two Layer Stochastic Interception Model, allows further insights. This model, which takes into

Table 5.3 *Interception and transpiration observations for forest, heather and grass summarized in terms of the average interception ratio α, the daily interception model parameters, γ, δ, and the ratio of actual to Penman E_T evaporation, β*

Site	Period	Interception Parameters			Transpiration fraction (β)
		α	γ (mm)	δ (mm^{-1})	
Forest					
All sites interception: (Plynlimon, Dolydd, Crinan and Aviemore)		0.35	6.9	0.099	
Plynlimon Forest Lysimeter	1974–1976	0.3	6.1	0.099	0.9
Dolydd	1981–1983	0.39	7.6	0.099	–
Crinan	1978–1980	0.36	6.6	0.099	–
Aviemore	1982–1984*	0.45	7.1	0.099	–
Heather					
Model estimate derived using automatic weather station data and measured interception parameters	1981	–	2.65	0.36	
Crinan, neutron probe	1981–1983	–	–	—	0.58–0.67
Law's heather lysimeters	1964–1968	0.16	–	–	0.25–0.5
Sneaton Moor lysimeter	1980	0.19	–	—	0.25–0.5
Grass					
Wye catchment, Plynlimon (indicates annual evaporation consistent with Penman E_T)					1.0

Note: *Not including snow periods.

Source: Calder (1990)

account the dependence of the rate of wetting of vegetation on drop size, indicates that the γ term will primarily be influenced by local climate and the aerodynamic and wetting properties of the canopy. The δ term will be controlled mainly by the wetting properties of the canopy. It is expected that from the knowledge gained from the further development and calibration of the more physically based interception models, it will eventually be possible to better identify the model parameters for use in operational evaporation models. These may take the form of the Two Parameter Exponential model or a 'collapsed' or simplified version of this model.

The seasonal model, as for the annual model, is appropriate for conditions similar to those of the UK uplands where soil moisture stress is an infrequent occurrence. The model, as described, is strictly applicable only to mature stands of vegetation. For catchments with a high proportion of immature

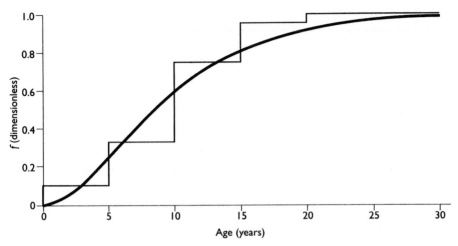

Figure 5.8 *'Effective' forest cover against age function*

forest it was suggested that, as a working hypothesis, the fractional canopy coverage parameter (f) could be related to age on the basis of an S-shaped function originally postulated by Binns (in Calder,1990, p122). Suggested values for f for upland UK forests were $f = 0.1$ for trees aged 0–5 years, 0.33 (6–10 years.), 0.75 (11–15 years), 0.95 (16–20years) and 1.0 for trees older than 20 years (Figure 5.8).

The LUC97 Model

From these studies, evolved the hydrological Land Use Change model (LUC), which now exists in a spatially distributed form that allows the convolution of rainfall and climate patterns with patterns of land use through linkage to a GIS. The LUC model also incorporates additional components, which take into account the effects of limiting soil water availability on transpiration (not usually required in the wet uplands of the UK). It also includes an improved and more general method for estimating the w term, a growth function that allows the treatment of immature forests and an option to allow seasonal variability in the interception parameters.

Limiting Soil Water Availability
The methodology for determining the effects on transpiration of soil moisture limitations follows the modelling approach used to describe the soil moisture regime under grassland sites (Calder et al, 1983) which used a moderating function (m) which limits as soil moisture is depleted, according to the relationship:

$$m = 1 \qquad\qquad (for\ \delta s < \tfrac{s_m}{2})$$

$$m = 2\left(1 - \frac{\delta_s}{s_m}\right) \qquad (for\ \delta s \geq \tfrac{s_m}{2})$$

where δs is the soil moisture deficit and the parameter s_m, is the maximum available water and represents the asymptotic value towards which the soil moisture deficit approaches. (It may also be regarded as approximately the total water available in the profile to the crop between the 'field capacity' and 'wilting point' values.)

Wet Day Fraction (w)

The fraction of the day vegetation canopies remaining wet, during and following rainfall, was estimated simply from the ratio of the daily interception loss and the maximum daily interception loss, for the particular vegetation type, using the relationship:

$$w = \gamma(1-exp(-\delta P))/\gamma$$
$$= 1-exp(-\delta P)$$

Growth Function

To take into account the hydrological response of the forest in the LUC97 model in the early, immature phase of the forest cycle, a one-parameter function, incorporating exponential terms has been used to describe the Binns sigmoidal function (Figure 5.8). The 'effective' fractional canopy coverage cover from fully-forested plots is described by:

$$f = 1-exp(-g_f a + 1 - exp(-g_f a))$$

where:
f is the 'effective' fractional canopy coverage
g_f is a 'growth factor', taken as 0.00047 (d^{-1});
and a is the age of the forest (d).

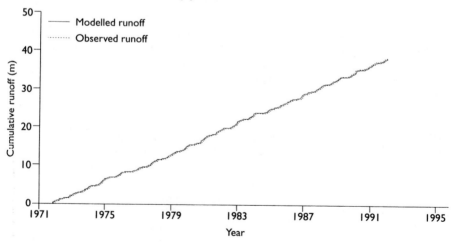

Note: Includes periods with partial felling of the forest.

Figure 5.9 *Cumulative runoff observed and predicted with the GIS version of the LUC model for the forested Severn catchment at Plynlimon*

Following the reasoning used in the original derivation of the model, the 'effective' fractional canopy coverage cover over the whole catchment can then be calculated from:

$$f = f_a(1 - exp(-g_f a + 1 - exp(-g_f a))) u^j$$

where
f is the 'effective' fractional canopy coverage of the catchment;
f_a is the fraction of the catchment area forested (fraction designated on a map as forest);
$(1 - u_j)$ is the fraction of the forested area occupied by road and river channel borders and gaps in the forest, taken as one-sixth. (This compares with the value of one-third which was used in the earlier model when immature forest within the overall forested area was also included).

Seasonal Variability
The present LUC97 version of the model also incorporates an option to allow seasonal variability in the interception parameters, which is particularly relevant to the treatment of deciduous forests.

Upland Forest Impact Predictions

Application of the GIS-linked versions of the LUC models shows that they are able to describe changes in flow resulting from afforestation practices in both the Plynlimon experimental catchments (Figure 5.9) and for river catchments in Scotland (Figure 5.10).[1]

The LUC97 model has also been applied to lowland forests in the UK, upland forests in New Zealand and variants of the model have been applied in Malawi and the Zambezi basin as described elsewhere in this chapter.

Trees and Drought: Lowland Conflicts

Quantity versus Quality

Among the recommendations of the UK government's 1995 White Paper on

---

1 The use of these models also has relevance for the quality control of data. Application of these models at Plynlimon helped to reveal inadequacies in the raingauge networks and inadequacies in the procedures used to analyse the data from these networks. These inadequacies include the use of data from ground level storage gauges in snow conditions (rather than daily measurements of snow depth and snow water equivalent), persistent leakage from some ground level gauges over a period of years and the use of canopy-level gauges for estimating rainfall inputs in the forested catchment, (including their use during snow periods), and the lack of a cosine correction to take account of slope of ground level for gauges inserted parallel to the slope. When these inadequacies are addressed, the LUC97 model gives good agreement with the observed flow regimes from both the grassland (Wye) and the partially forested (Severn) catchments at Plynlimon. Furthermore, when these inadequacies are addressed, there is no indication of a reduction in the evaporation from the Wye catchment with time, that had earlier been claimed by Hudson and Gilman (1993), nor is there any indication of any reduction in the evaporation from the Severn catchment which cannot be explained by changes in forest area resulting from felling operations.

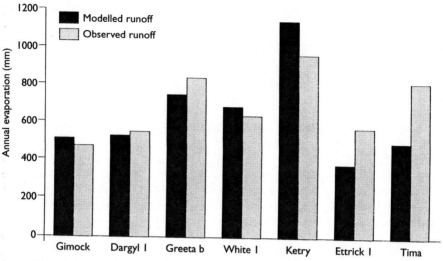

Source: Price et al (1995)

Figure 5.10 *Annual evaporation, measured and observed, from seven upland forested catchments in the UK*

Rural England (DOE, 1995) was a proposal to double the area of forests within England by the year 2045. The combination of this proposal together with government recognition of the real possibility of climate change resulting in hotter, drier summers and wetter winters, has raised questions (House of Commons, Environment Committee, 1996; DOE, 1997) concerning the possible impacts on UK water resources and the water environment of the combined effects of climate change and such a large change in land use.

Interactions between water quality and quantity were also recognized as an issue. Woodlands can protect water supplies from nitrate pollution associated with agriculture and this benefit needs to be balanced against any reduction in water yield. Nitrate concentrations within UK groundwaters have been steadily rising over the last 20 years due to leaching losses from intensive agriculture. This trend is expected to continue for at least the next 10 to 15 years. Many sources, particularly those within the Triassic sandstone aquifer in the UK midlands, are now close to, or exceed, the mandatory 50 mg/l standard for potable water. The protection afforded by forests has recently been recognized and new boreholes have been sunk within existing forest on the sandstone aquifer to tap the low nitrate water (nitrate concentrations less than 10 mg/l) for blending purposes. The development of such sources could avoid the need for expensive nitrate removal treatment.

Clearly to be able to follow the philosophy of IWRM it is important to be able to assess the water quantity impacts of lowland afforestation both now, and in the future when climate change may have occurred, and to set these within the context of other water quality, environmental, economic and socio-economic impacts. Unfortunately, it is difficult to make an accurate prediction of the water quantity impacts of UK lowland afforestation, even under the present climate, for two reasons.

The first reason is that in the lowlands of the UK, both the effects of higher interception losses and greater transpiration during dry periods from forests operate, but neither predominates. Furthermore, in non-rainfall periods, but periods when soils are sufficiently wet to offer no restriction to the availability of soil water to either short crops or forests, it is usually found that forest transpiration rates are actually ~10 per cent less than those from the shorter crops. As processes are at work that can either increase or decrease forest evaporation the prediction of forest, as compared with short crop, evaporation in the present British lowland climate becomes very uncertain.

Secondly the information on the evaporative differences of combinations of different tree species growing on different soil types in the lowlands of the UK is limited. For some important tree species and soil type combinations, it is non-existent. Virtually no information is available on the evaporative characteristics of trees growing on soils overlying sandstone, or for that matter on "brownfield sites", yet it is expected that much of the new planting will take place in the Midlands of England on just these types of soil. There is also doubt whether information, which is available for particular tree species and soil types, is correct. This is particularly the case for broadleaf forest on chalk soils so it may not always be possible, using existing information, to determine the direction of the impact let alone the magnitude.

Earlier research illustrates the difficulties. One important study, carried out at Thetford forest in the east of England in the early 1980s (Cooper, 1980) showed that recharge under the pine forest was reduced by as much as 50 per cent as compared with grassland. Here the trees were able to tap water from the underlying chalk whereas the grass, through shorter rooting, had limited access to water stored in the overlying sandy soil. A second study (Harding et al, 1992), at Black Wood in southern England, referred to in Chapter 3, concluded that beech afforestation of grassland overlying the chalk would increase recharge by 18 per cent. A third study (Hall et al, 1996b), commissioned by the Department of Trade and Industry (DTI) to investigate the water use of fast-growing coppice poplar and willow plantations in southern England, growing on clay soils, indicates evaporative differences similar in magnitude to the Thetford results.

The Black Wood Anomaly

Published results on the water use of forest in lowland England indicate a range of impacts, with the Black Wood results alone indicating increases in runoff or recharge as a result of afforestation. Although the measurements of soil moisture depletion recorded in the chalk soils beneath the beech forest at the Black Wood site in southern England are accepted as being reliable (see Chapter 3 and Figure 5.11), it has been questioned (Calder et al, 1997b) whether these measurements necessarily support the hypothesis that beech-forest water-use is less than that from grassland. This is an important question because the interpretation given by Harding et al (1992) has left the UK water industry with the sanguine view that broadleaf afforestation of grassland areas will be of positive benefit to water resources, a view that is, to say the least, unusual in the world context.

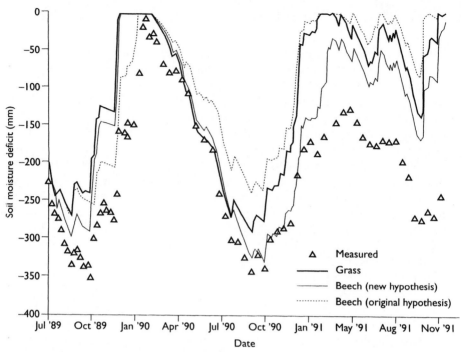

Figure 5.11 *Model predictions of soil moisture deficits due to evaporation at the Black Wood site for beech (original and new hypotheses) and grass together with the observed soil moisture deficits under beech*

The earlier conclusions drawn concerning the evaporative losses from forest at the Black Wood site were partly based on assumptions of very significant drainage (~150 mm) taking place from the soil/chalk profile during soil moisture deficit conditions that was required for consistency with transpiration calculated using measured stomatal conductance. To investigate whether alternative interpretations were possible for the Black Wood results, a 'limits'-type model, a version of LUC97, was used to explore different hypotheses. The hypotheses investigated were whether the original assumption made about the drainage function for the chalk or whether a new function, implying much reduced drainage (~25mm) (Calder et al, 1997b), was correct. The version of LUC97 was standard except for the use of an exponential 'step length' function (Calder et al, 1983), to represent freely available water and to model soil moisture deficits in relation to soil moisture availability, to be consistent with the approach used by Harding et al (1992).

The LUC97 model predictions of soil moisture deficits due to evaporation alone (without incorporating a drainage function) obtained with model parameters relating to the original hypothesis and the new hypothesis, are shown in Figure 5.11. For comparison the measured SMDs are also shown in the same figure. The model parameters relating to the original hypothesis (Harding et al, 1992) and the new hypothesis, which were obtained partly from 'default' values (Calder, 1990) and partly by adjusting the interception model parameters to give a better fit to the observed SMD, are given in Table 5.4.

Table 5.4 *Evaporation model parameters representing different vegetation covers on different soils where step length relates to available soil water, β is the transpiration fraction and γ and δ are interception parameters*

Chalk	Grass		Beech (original hypothesis)	Beech (new hypothesis)
Source	Calder et al (1983)		Harding et al (1992)	Calder et al (1997b)
Step Length	160		1000	1000
β	1		0.75	0.9
γ	0		2.23	4.46
δ	–		0.21	0.099
Winter γ	0		1.84	3.68
Winter δ	–		0.108	0.099

Sand	Grass	Pine	Broadleaf (new hypothesis)	Mixed forest
Source	Calder et al (1983)	Cooper & Kinniburgh (1993)	Cooper & Kinniburgh (1993)	
Step Length	53	83	83	83
β	1	0.9	0.9	0.9
γ	0	4.6	4.46	4.6
δ	–	0.099	0.099	0.099
Winter γ	0	4.6	3.68	3.68
Winter δ	–	0.099	0.099	0.099

Clay loam	Grass	Pine	Broadleaf (new hypothesis)	Mixed forest
Source	Calder et al (1983)	Cooper & Kinniburgh (1993)	Cooper & Kinniburgh (1993)	
Step Length	75	200	200	200
β	1	0.9	0.9	0.9
γ	0	4.6	4.46	4.6
δ	-	0.099	0.099	0.099
Winter γ	0	4.6	3.68	3.68
Winter δ	-	0.099	0.099	0.099

Source: Calder et al (1997b)

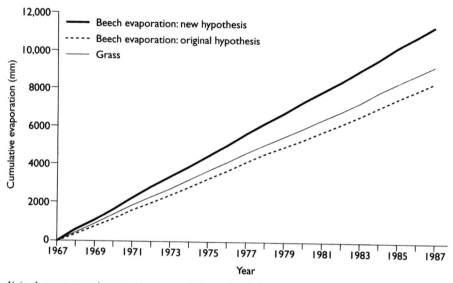

Note: Average annual evaporation: grass, 468 mm; beech (original hypothesis), 424 mm; beech (new hypothesis), 573 mm; average annual rainfall: 741 mm.

Figure 5.12 *Predicted cumulative evaporation for different land uses at the Black Wood chalk site (1967–1987)*

Model predictions of the cumulative evaporation from grass and beech forest, assuming both hypotheses, are shown in Figure 5.12. The new hypothesis would indicate that, as a long-term average, the annual evaporation from broadleaf forest (beech) is 105 mm higher than that from grassland. The average recharge (assuming no runoff from the chalk site) would be reduced by 38 per cent as a result of broadleaf afforestation of grassland. This predicted reduction of recharge of 38 per cent should be seen in contrast to the increase in recharge of 15 per cent predicted by the earlier Black Wood study.

Forest Impacts in the Midlands

Following the original study of Harding et al (1992) the models developed at the Black Wood chalk site were extended to estimate the impacts of afforestation on sites overlying the Nottingham Triassic sandstone, one of the most important aquifers in the UK (Cooper and Kinniburgh, 1993). As Cooper and Kinniburgh used the same model parameters to describe the evaporative response of broadleaf forest at the Nottingham sites, it is not surprising that they drew similar conclusions: that broadleaf afforestation of grassland on the Nottingham Triassic sandstone would also increase recharge.

Nevertheless, if the reservations concerning the use of these parameters to describe the evaporative response of forest on chalk are well founded, the same reservations must apply to the conclusions drawn of reduced evaporation from broadleaf forest, as compared with grass, on sandstone. To investigate the range of possible impacts resulting from broadleaf afforestation on sandstone sites, a similar modelling study to that outlined above for

Greenwood Community Forest

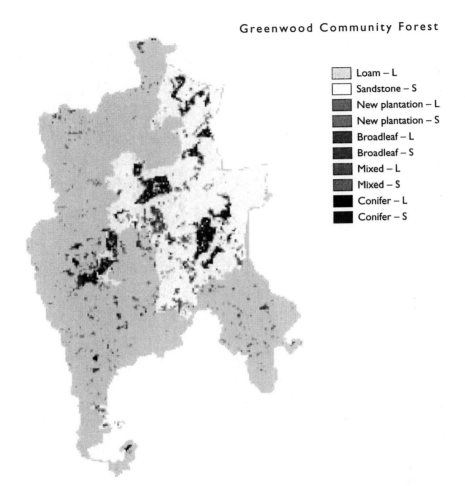

Loam – L
Sandstone – S
New plantation – L
New plantation – S
Broadleaf – L
Broadleaf – S
Mixed – L
Mixed – S
Conifer – L
Conifer – S

Figure 5.13 *GIS display of forest cover in the Greenwood Community Forest*

Black Wood, was carried out. This study used essentially the same new hypothesis model parameters derived for broadleaf forest, but adjusted it to take into account the different soil water availability expected on sand and clay-loam soils in the Midlands. The chosen study area was the Greenwood Community Forest in Nottinghamshire. Nottinghamshire County Council supplied land use information, relating to forest cover and the distribution of agriculture and grassland together with information relating to the geology and soil type on GIS files (Figure 5.13).

Application of LUC 97 then allowed the calculation of the range of impacts associated with the two scenarios (Calder et al, 1997b).

The model predictions of seasonal evaporation, assuming essentially the same model 'new hypothesis' parameters relating to the chalk site, but with different parameter values relating to soil water availability (Table 5.4), are shown in Figure 5.14.

The new hypothesis assumption would indicate that as a long-term average, the annual evaporation from broadleaf forest on sand soils is 93 mm

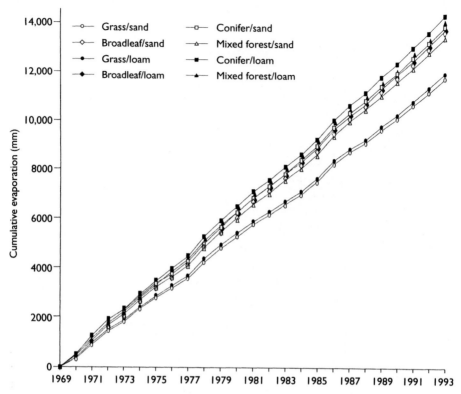

Note: Average annual evaporation: grass/sand 460 mm; conifer/sand 571 mm; broadleaf/sand 553 mm; mixed forest/sand 554 mm; grass/loam 486 mm; conifer/loam 603 mm; broadleaf/loam 594 mm; mixed forest/loam 594 mm. Average annual rainfall: 628 mm.

Figure 5.14 *Predicted cumulative evaporation for different land uses at the Greenwood Community Forest (1969–1993)*

higher than that from grassland, and that the average recharge plus runoff would be reduced by 51 per cent as a result of broadleaf afforestation of grassland. For broadleaf afforestation on clay-loam soils, the predicted reduction in recharge plus runoff would be 62 per cent.

The calculated cumulative recharge plus runoff from the Greenwood Community Forest assuming the present forest cover is shown in Figure 5.15. Also shown is the calculated cumulative recharge plus runoff for a threefold increase in forestry, where it is assumed that the increase occurs in proportion to the present distribution of forestry on the different soil types. Over the 24-year period from 1969 to 1993, the calculated average reduction in recharge plus runoff from the Forest, as a result of an increase in forest cover from the existing 9 per cent to 27 per cent, is 14 mm (11 per cent reduction).

The Unresolved Question

The model predictions obtained assuming the different hypotheses indicate hugely different projections with regard to the hydrological impact of

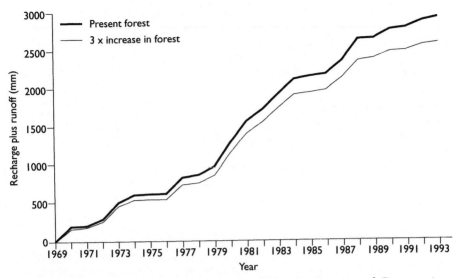

Figure 5.15 *Calculated recharge plus runoff for the Greenwood Community Forest assuming present forest cover throughout the period 1969–1993 and for a threefold increase in forest cover*

increased lowland broadleaf forests. Earlier studies had indicated that broadleaf afforestation of grassland overlying chalk or sandstone bedrock would have beneficial impacts on water resources by increasing recharge. The alternative scenario indicates that from both a national and local perspective, the implications could be very significant. Of primary concern, if this scenario is correct, are the local implications of increased forestry in areas where water resources are already being used to the limit or where low flows in rivers are causing environmental concerns. Further research is planned to determine which of these scenarios is correct.

NEW ZEALAND: WATER ISSUES AND LAND USE

The Perception of Water Issues

On any worldwide comparison, New Zealand must rate high as a country with abundant water resources. Twelve metres or more of water fall each year on parts of the Southern Alps in the South Island. Nor, with only 3.4 million people occupying a land area of 270,000 km², is the population pressure high. Taken together with a low-pollution climate and generally high-quality waters, it would be easy to assume that water resource issues were not a major concern in New Zealand and that water resource conflicts would be minimal.

Yet this sanguine view is not wholly true. Although abundant, New Zealand's water resources are not well distributed. Average annual rainfall at Milford Sound on the west coast of the South Island is 6240 mm, but within a distance of not much more than 100 km eastwards, at Alexandra in central

Table 5.5 *Estimated economic value of New Zealand's water resources*

Activity	Value (NZ$ million)
Water supply (agriculture, industry, domestic)	450
Waste disposal	450
Freshwater fisheries	100
Recreation and amenity values	500
Hydroelectric power generation and thermal plant cooling	800
Gravel resource replenishment	40
Total	2340

Source: Mosely (1988)

Otago, the rainfall is only 340 mm. The eastern areas of both islands have dry summers and experience soil moisture deficits sufficient to limit horticultural development. Hydroelectric power stations supply a major part of New Zealand's power requirements, but in the South Island much of the rainfall on catchments falls in summer whilst electricity demand peaks in winter. In these high-flow rivers, floods are a common hazard and high flows are commensurate with high rates of sediment transport. Water quality has been affected in some catchments by effluents from urban areas and facilities such as dairy and wood processing plants, by runoff enriched by fertilizer and animal wastes from agricultural areas, and by sediment introduced by accelerated erosion (Waugh, 1992). Water-based recreation, boating and fishing are popular pastimes in New Zealand and the amenity aspects of high quality waters are appreciated. As in so many countries, irrigation, which uses 1.1 $km^3.yr^{-1}$, is the largest consumptive user of water, but is not the highest value use (Table 5.5).

Waugh (1992) predicts an increasing conflict between water conservation interests and farmers needing irrigation water to support pastoral agriculture, cropping and horticulture in New Zealand.

From a world perspective these hydrological concerns might seem neither unusual for a developed country nor particularly severe. For cultural and traditional reasons this is not how the issues are perceived in New Zealand, particularly in relation to waste disposal. Whereas the European settlers, following on the tradition of European settlers elsewhere, showed no particular appreciation for the environment and were happy to throw their wastes and effluents into watercourses, this is not a practice that the Maori inhabitants would condone or accept. The 'new' concept of sustainability is not new to the Maori who regards water as the essential ingredient of life – a priceless treasure left by ancestors for the life-sustaining use of their descendants (Taylor and Patrick, 1987).

In the early 1950s concerns about the impacts of land use change and upland land use on water yield and other aspects of hydrology led to New Zealand embarking on a major programme of land-use change catchment studies to quantify the impacts. In many ways the issues and the impacts are not too dissimilar to those in the UK and similar methods have been used to research the impacts. The modelling methodologies used to describe and predict the impacts of land use change on water yield are outlined below.

Land Use Change and Impacts

Major changes in New Zealand's land cover followed the Polynesian settlements about 1000 years ago. Prior to the first settlement, forest is thought to have covered 75 per cent of the 26.5 million hectares of land. By the time the Europeans arrived only about 11.3 million hectares of forest remained and by 1950 the area of native forest had been reduced to about 5.7 million hectares, much of it on the poorer, mountainous land. About the same area of land remains under native tussock grass (Fahey and Rowe, 1992). Planting of exotic forests over large areas began in the 1920s and by 1990, 1.3 million hectares had been established. About 9 million hectares are under improved pasture. The country's economic growth was dependent on these land use changes, but it was soon recognized that these alterations of land use were having important impacts on the hydrology particularly in relation to the quantity of water, floods and erosion.

The land use impacts section of the Forest Research Institute (now Landcare Research New Zealand Ltd) operates three paired catchment experiments in the South Island. These are at Maimi, to investigate the impacts of harvesting native beech forest; at Big Bush to determine the impacts of establishing *Pinus radiata* plantations; and at Glendhu, to determine the impacts of converting native tussock grassland to pine. To illustrate how land use change methodologies have a general application in helping with the resolution of water resource conflicts, the use and application of the LUC97 model largely developed and calibrated for use in UK conditions, to New Zealand conditions, is described below.

The Glendhu Study

Proposals to convert large areas of tussock grassland in the upland of east Otago to pine plantation have raised concerns about the impacts on both water quantity and quality. In 1979 the former New Zealand Forest Service established a paired catchment study in the Glendhu forest in the upper Waipori catchment, where water quantity impacts were a major concern to hydroelectric generating plants located downstream. The aims were to investigate the streamflow behaviour and water balance of lightly-grazed tussock grassland and to assess the impact of afforestation on annual water yields, storm peak flows, water chemistry and sediment yield. Only the impacts on yield are described here.

The Glendhu paired catchment experiment was instrumented in 1979 with rain gauges and flow-measuring weirs: data collection began on 10 October 1979 (Fahey and Watson, 1991; Fahey and Jackson, 1997). The two catchments, GH1 and GH2, were originally both under native tussock grass-

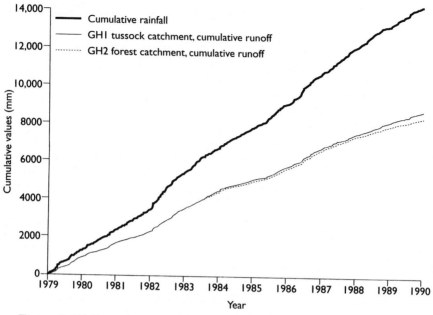

Figure 5.16 *Cumulative rainfall and runoff recorded from the Glendhu catchments*

land (dominated by the species *Chionochloa rigida*). In 1982, after a three-year calibration period, 67 per cent of GH2 was planted with *Pinus radiata*. Catchment GH1 was retained under the original vegetation as the control catchment.

Changes to the Flow Regime

The cumulative rainfall and flow record from the catchments is shown in Figure 5.16.

The very close agreement between the runoffs measured from the two catchments during the calibration period implies a very high degree of integrity and internal consistency of the data from these catchment studies. From 1987 onwards it can be seen that the cumulative flow recorded from GH2, the forested catchment, is less than that from the control, GH1, and that the divergence becomes increasingly pronounced as time progresses. The difference in slopes in the more recent period, 1991–1994, indicates a reduction in runoff of 27 per cent as compared with the control.

Application of the LUC97 Model to Glendhu

Application of the LUC model to the Glendhu tussock-grass catchment showed that good agreement between predicted and observed runoff could be achieved with parameter values only slightly modified from those measured for heather moorland containing *Calluna vulgaris* (Calder, 1990) in the uplands of the UK (Table 5.6).

Table 5.6 *Evaporation model (LUC97) parameters representing tussock and pine vegetation on soils with high water availability*

LUC97 parameter	Heather	Tussock grass	Pine
Source	Calder, 1990 (measured)	Calder, 1990 (based on heather parameters)	Calder, 1990 (identical to conifer parameters)
s_m	200	200	380
β	0.5	0.5	0.9
γ	2.65	2.4	6.91
δ	0.36	0.36	0.099
Winter γ	2.4	2.4	6.91
Winter δ	0.36	0.36	0.099

Note: Where s_m relates to maximum available soil water, β is the transpiration fraction and γ and δ are interception parameters.

A ten per cent reduction in the γ parameter, which determines the maximum daily interception loss, from 2.65 to 2.4 allows good agreement both seasonally and in the long term totals of flow (Figure 5.17).

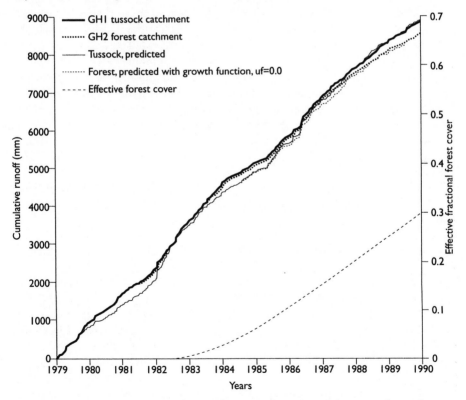

Figure 5.17 *Cumulative runoff measured and predicted for tussock catchment GH1, together with cumulative runoff measured and predicted for forest catchment GH2, using the forest cover/age function*

By incorporating the LUC97 standard upland forest growth function (see above) and the same parameter value derived for spruce plantations in the UK, the LUC97 model was also able to describe the change in the flow regime that was measured at Glendhu as the forest grew (Figure 5.17).

The modelling study confirms the significance of forest impacts on water yield and provides a modelling framework for estimating the impacts of afforestation in other upland areas of New Zealand.

THE PHILIPPINES: BIG CITIES, SMALL CATCHMENTS

The rapid industrialization and urbanization of the major population centres in the Philippines, together with logging, agricultural and irrigation developments in the headwater catchments, have induced environmental and water resource concerns which, in terms of both range and severity, exemplify many of the problems that face large cities in the developing world. Whilst increasing populations and expanding industries demand more water, the ability to meet demands is being reduced by siltation, nutrient, herbicide and pesticide pollution of surface waters, together with pollution and often saline intrusions within the groundwater aquifers.

The Philippines is a country which is made up of more than 7000 islands, and many of its provincial centres, including Metro Manila, are coastal. The economic boom of the early 1990s has resulted in rapid development which has often been to the disadvantage of the environment, so that air quality, water quality and quantity, degradation of the land and marine ecology are all issues of great national concern. But pressures for further development are leading to increased urbanization which is now moving from the coastal plains into the headwater catchments. Regulations on development are in the hands of many government departments (Tolentino, 1996) which have not always been successful in demonstrating an integrated approach to land and water resource planning.

The problems affecting the Philippines' second city, Cebu, are illustrative of these issues and a novel grass-roots approach to IWRM is described below.

Cebu – A City with Water Stress

Concerns about Cebu's water supplies were first voiced in the mid-1970s. The Water Resources Centre of the University of San Carlos in Cebu City detected an increasing intrusion of sea water into Cebu's aquifer system which was attributed to over extraction. By 1997 the water demand in Metro Cebu was estimated at 240,000 cubic metres per day of which only 115,000 cubic metres per day was being supplied by the government-owned and controlled water supply corporation. The deficit was being met by uncontrolled extractions from private wells whilst industrial needs were being met by a combination of water purification and desalination.

The land use on the three headwater catchments which supply surface water to Cebu was another cause for concern. The original forest cover on these catchments had been cleared for timber products and agricultural activities which had seriously degraded water quality. Efforts to improve land management on the catchments had been hindered by questions of land ownership between land owners, tenant farmers and occupiers. Questions of jurisdiction between local government units and national government agencies had also confounded the problem.

An Asian Institute of Management report (Tañada, 1997) claimed:

> '*The initial response to the water crisis by the local government units was slow and tentative in some cases. Towns and cities in Metro Cebu had few specific programs in informing the local residents about the water situation. Frequently, local officials, including mayors, councillors, development council members, as well as local line agency personnel had little or no knowledge about water resources problems, let alone strategies for solving them. They did not have access to relevant information and frequently made decisions based mostly on political and/or economic considerations without regard for their long term consequences. Their decisions were also not based on a holistic approach.*'

The forestry lobby advocates reafforestation with fast growing tree species as the panacea, in the fond belief that this will automatically cure all water quality and quantity problems. Tañada (ibid) says, with perhaps some understatement,

> '*the challenge of finding a common point for cooperation or compromise between the various groups is not an easy task. The absence of an integrated management of resources and comprehensive land use policy for the watersheds has left the area open to unregulated and inappropriate land use.*'

Cebu Uniting for Sustainable Water – a Citizens' Initiative for IWRM

The perceived failure of governmental planning has resulted in a pressure group, Cebu Uniting for Sustainable Water (CUSW) being set up to address these land and water resource issues (Tañada, ibid). Formed out of a coalition of several organizations in Metro Cebu, CUSW regards itself as a citizens' initiative. Since its formation in 1995 it has been working towards developing a land-use plan for the watersheds. It realizes that the real test of its success will be whether the plan is eventually adopted by the leaders of Metro Cebu.

At its general assembly in April 1997, CUSW had 77 institutional and 66 individual members covering many sectors. With such a range of public and private, socially and business-oriented interests represented (Table 5.7), CUSW encountered dissenting opinions, not only outside but also within the group.

Table 5.7 *Sectors represented by 'Cebu Uniting for Sustainable Water'*

Sectors represented by CUSW			
Farmers	Fisherfolk	Business	Professionals
Youth	Academia	Religious	Political leaders
	Local government	Landowners	Hill-land residents
NGOs	Line agencies	Women	Health
Urban poor	Cooperatives	Civic	Labour
Interested individuals			

Source: Nacario-Castro (1997)

CUSW claims not to take a hard line against development nor claims to be an environmental group. It claims to be seeking schemes which allow development, but which are compatible with maintaining the watershed so that the supply and quality of the water are enhanced. CUSW recognizes that the interests of the people who live in the area and survive from the land have to be considered – displacement is not feasible, socially or economically. The eleven objectives that CUSW hope to see incorporated in the Cebu Watershed Management and Development Plan (Nacario-Castro, 1997) are :

1   Integration of watershed communities.
2   Respect for prior right over alienable and disposable lands.
3   Protection and promotion of biodiversity.
4   Adherence to a watershed resource management plan.
5   Use of biodegradable and non-toxic substances.
6   Installation of control systems against potable water contamination, atmospheric pollution and noise reduction.
7   Respect for the dynamism of local culture.
8   Promotion of human security agenda.
9   Incorporation of a method for measuring social impact.
10 Non-discrimination of local residents.
11 Extended accountability of project developers.

Cebu – Further IWRM initiatives

Under the sponsorship of the British Council, a Watershed Protection and Management Seminar was held in Cebu in November 1997. The seminar was attended by all sectors concerned with the management of land and water resources in the Philippines and the output of the seminar was a statement subscribed to by the organizations represented. This statement, which is consistent with the stated objectives of the CUSW Watershed Management and Developed Plan, is a framework to improve watershed management.

The required components of this framework are:

1   National Watershed Management body with operating units at watershed level.
2   Integrated Watershed Management plan including land use.
3   Survey, studies, data-based planning, consultation, participation on the part of stakeholders.
4   Data and decision support system.
5   Development of management programmes.
6   Programme evaluation and assessment.
7   Monitoring and promulgation.
8   Resources: financial and technical.
9   Commitment, vision, and understanding of strategy process.

Implementation programmes are as follows:

1   Legislation of ordinances for watershed management including land tenure, water rights.
2   Education, training and workshops for stakeholders.
3   Establishment of national authority.
4   Body for enforcement, implementation, monitoring and facilitation.
5   Establishment of decision support system.
6   Creation of water information system.
7   Control pollution.
8   Appropriate technology.
9   Water pricing policy, costing, valuation of resources and fines.

Activities and delivery mechanisms consist of:

1   Pilot programme on small basin.
2   Fund sourcing.
3   Coordination among groups.
4   Legislation and policy.
5   Community participation.
6   Capacity building and technical support.
7   Provision of livelihood opportunities.
8   Education (proactive), dissemination – all stakeholders.
9   Networking – all stakeholders.
10 Coordinating body.
11 Review existing process.
12 Development of an implementation plan.

SOUTH AFRICA: FORESTS AND WATER

The Revolution – Civil Rights and Water Policy

> *'The dictionary describes water as colourless, tasteless and odourless – its most important property being its ability to dissolve other substances. We in South Africa do not see water that way. For us water is a basic human right, water is the origin of all things – the giver of life.'*
> Source: South African White Paper on Water Policy

For a country which has so courageously faced and resolved inequities in its attitudes towards, and its political treatment of, different racial groups, it would perhaps not be surprising if South Africa were prepared to exhibit the same virtues in dealing with its land use and water resource issues. It might be expected that under such circumstances South Africa's White Paper on Water Policy would display new approaches to dealing with the issues. In the event the White Paper demonstrates such remarkable vision and awareness of the issues that it must be regarded as the model for other countries to follow and must place South Africa at the very forefront of the blue revolution.

Both Africa's spiritual and secular appreciation of water is well captured in the poems taken from the White Paper:

> *'From water is born all peoples of the earth. There is water within us, let there be water with us. Water never rests. When flowing above, it causes rain and dew. When flowing below it forms streams and rivers. If a way is made for it, it flows along that path. And we want to make that path. We want the water of this country to flow out into a network – reaching every individual – saying: here is this water, for you. Take it; cherish it as affirming your human dignity; nourish your humanity. With water we will wash away the past, we will from now on ever be bounded by the blessing of water.'*
> Source: Mazisi Kunene

> *'Water has many forms and many voices. Unhonoured, keeping its seasons and rages, its rhythms and trickles, water is there in the nursery bedroom; water is there in the apricot tree shading the backyard, water is in the smell of grapes on an autumn plate, water is there in the small white intimacy of washing underwear. Water – gathered and stored since the beginning of time in layers of granite and rock, in the embrace of dams, the ribbons of rivers – will one day, unheralded, modestly, easily, simply flow out to every South African who turns a tap. That is my dream.'*
> Source: Antjie Krog

South Africa has the political will to address water, land use and forestry issues and a governmental structure which is almost uniquely suited for the purpose: both the water and forestry sectors are within the same ministry, the Department of Water Affairs and Forestry (DWAF). To achieve the goal of *'Equity, efficiency and sustainability in the supply and use of water in South Africa'*, Professor Kader Asmal, Minister of DWAF, has initiated the National Water Conservation Campaign and conservation projects which include the Working for Water Programme (DWAF, 1996) (see below).

South Africa has also carried through a programme of adaptive forest hydrology research that has made use of knowledge gained in other parts of the world and supplemented this with research on forest-water interactions in South African conditions. This carefully executed research, carried out in close collaboration with stakeholders both with forestry interests and downstream user and environmental interests, has resulted in findings which are not only widely disseminated but are also accepted by these different groups.

This has led South Africa not only to endorse the conventional 'polluter pays' principle which requires the polluter of the environment to either pay for its remedial treatment or to pay society in recompense for the loss of environmental quality, but to originate a new principle, the 'user pays' principle. This requires land uses that consume large amounts of water, such as forestry, to pay what is now commonly becoming known as an 'interception levy'. South Africa is unusual in having the awareness and confidence to address the issues and the confidence in its research to advance this new approach.

Water and Land Use Issues

Water has always been a major concern in South Africa. The average annual rainfall amounts to 440 mm of which less than ten per cent reaches the rivers. In the 1920s, farmers' associations and other organizations petitioned the government to investigate why many of South Africa's rivers were drying up. At the time there was a drive to encourage tree planting. The forester's myth of the hydrological benefits of forests was under suspicion. There were concerns that extensive plantations of exotic pine, eucalypts and wattle were reducing water supplies, exhausting the soil and promoting erosion. The issue of forestry and water supplies was discussed at the Fourth Empire Forestry Conference hosted by South Africa in 1935 and as a result a research station was established at Jonkershoek in the south-western Cape with the task: *'to determine how normal afforestation, as carried out in State plantations, would affect climate, water conservation and erosion'*. Dr C L Wicht, founder of the station, devised an experimental design of catchment research which was based on the paired-catchment principle as used at Emmental in Switzerland and Wagon Wheel Gap in Colorado, USA. But Dr Wicht was quick to realize that no two catchments of the eight identified for research at Jonkershoek were even remotely comparable. He devised a novel approach:

'Each stream is to be studied independently and compared with itself before and after treatment. In each case all factors which might influence streamflow will be observed and correlated. It is hoped that such analysis will disclose general trends common to all catchments. On the basis of these trends it may be possible to generalize as to the effects of afforestation on streamflow in the winter rainfall region.' (Wicht, 1939)

It was this attention to detail that ensured that the potential of the catchment study approach was maximized so that any identified changes in the flow regime could be correctly attributed to either the change in land use or to other climatic or environmental variables. This 'belt and braces' approach entailed extra costs in the operation of both the calibration period and the paired catchment comparison period, and delays in obtaining the final result (because of the extra time required for the calibration). With hindsight, this was clearly justified by the achievement of unambiguous and indisputable results: results which, because they are so clear-cut and unmistakable, have been accepted by both the scientific and policy-maker communities. Since 1972, the Forest Act has required timber growers to apply for permits to establish commercial plantations on new land or sections of land which have not been planted for five years, and applications may be rejected on the grounds of high water use. Robust empirical models now exist (Scott and Smith, 1997) for calculating the reduction in total and low flows to assist with this task. This acceptance of the findings of the carefully conducted research in South Africa has led to them being used as the basis of water and land management policy. In other countries discussion and debate is still clouded by the 'forester's myth' and by short term and sometimes poorly conducted research which has given ambiguous results. This research has also led to the question 'should South Africa, a water-scarce country, be exporting its water in the form of forests?'

Upland Water and Land Use Conflicts

Mountains are the dominant influence on South Africa's surface water resources. Only 20 per cent of the country receives more than 800 mm of rain and most of this is in mountainous areas. For such a water-deficient country (with an average annual rainfall over the whole country of only 400 mm) the provision of water from the uplands, maximal in quantity and quality, is of vital concern.

The forests require at least 800 mm of rain to grow at economic rates so these same mountain catchments come under pressure both for afforestation and for water gathering. Commercial forests consist almost entirely of exotic species and form a large and important industry with plantations occupying almost 1.2 million hectares (Bands et al, 1987). Nowhere are the conflicts between forestry and water interests more extreme than in South Africa.

The predominant indigenous vegetation cover in the mountains is not forest. Under the influence of periodic bushfires, grassland and unique *fynbos* communities have evolved as the natural vegetation cover, with patches of indigenous forest being confined to small areas on cool and protected sites. Together with demands for water and economic driven demands for timber, there are now ecological demands for the *fynbos* to be conserved.

Upland Land Use, Ecology and Hydrology Interrelationships

Understanding of the interrelationships between land use, hydrology and ecology is essential for the sustainable and multi-use management of South Africa's mountainous areas. The hydrological research programme centred at Jonkershoek, which by the 1960s had become even more focused on ecological concerns, is providing that understanding. The catchment studies provide unequivocal evidence for the reductions in streamflow that will occur as a result of afforestation with commercial species (Table 5.8).

The studies also destroy the forester's myth of forests 'attracting rain'.

> *'Forests are associated with high rainfall, cool slopes or moist areas. There is some evidence that, on a continental scale, forests may form part of a hydrological feedback loop with evaporation contributing to further rainfall. On the Southern African subcontinent, the moisture content of air masses is dominated by marine sources, and afforestation will have negligible influence on rainfall and macroclimates. The distribution of forests is a consequence of climate and soil conditions – not the reverse.'*
> *Source: Bands et al (1987)*

The studies also confirmed that periodic fire and its management were essential not only for the maintenance of the natural vegetation, but also for the maintenance of the soil mantle, the unique fauna and water yield. To maintain the ecology of the *fynbos*, fire is needed at intervals of between 10 and 30 years to germinate seedlings. Fire intervals shorter than about six to ten years eliminate many plant species, whilst protection from fire for more than 30 years results in senescence. Burning of the *fynbos* also increases water yield; the burning of 23 year-old *fynbos* resulted in streamflow increases of 200 mm in the first year after burning (Bands et al, 1987). Burning of grassland has a negligible effect on water yield because of the winter dormancy of grass and its rapid regrowth to full canopy in the spring.

A major threat to the *fynbos* ecology and to the water yield from these upland areas is from the invasion of alien shrubs and trees. Two of the most dangerous species are the Australian shrub, silky hakea (*Hakea sericea*), and the Mediterranean cluster pine (*Pinus pinaster*). A major programme of IWRM directed at eliminating or controlling these invaders, to conserve water yields and the ecology of these areas whilst providing economic returns and employment opportunities, is now underway.

Table 5.8 *Impact of afforestation on streamflow determined by catchment studies in South Africa*

Experimental catchment	Natural vegetation	Rainfall (mm)	Natural runoff (mm)	Afforestation	Streamflow reductions (mm)
Bosboukloof Jonkershoek	Fynbos	1300	600	*Pinus radiata* (57%) (1940)	330 at 23 years
Biesievlei Jonkershoek	Fynbos	1430	660	*P. radiata* (98%) (1948)	400 at 15 years
Tierkloof Jonkershoek	Fynbos	1800	1000	*P. radiata* (36%) (1956)	500 at 16 years
Lambrechtbos Jonkershoek	Fynbos	1500	530	*P. radiata* (84%) (1961)	170 (mean 8 to 16 years)
Catchment 2 Cathedral Peak	Grassland	1400	750	*Pinus patula* (75%) (1951)	375 at 17 years / 440 at 22 years
Catchment A Mokobulaan	Grassland	1150	250	*E. grandis* (100%) (1969)	403 at 5 years
Catchment B Mokobulaan	Grassland	1040	220	*P. patula* (100%) (1969)	100 at 5 years (tentative result)
Catchment C Mokobulaan	Grassland	1200	180	Control	–
Catchment D Westfalia	Indigenous forest	1700	720	*E. grandis*	200 at 3 years

Source: Bands et al (1987)

Invasive Trees

Probably nowhere else in the world is the tight link between land use and water resources better appreciated than in South Africa. Not only have research programmes been geared to determining the impacts of plantation forestry, for forest and land management purposes, but research is now being directed at the impacts and management of invasive species, some of which have 'escaped' from the commercial forests. These invaders have much the same impacts as forest plantations in reducing streamflow, but in the unmanaged state they also have other deleterious hydrological effects. The increased amount of plant material, the above-ground biomass, in invaded *fynbos* areas is three to ten times higher (Versfeld and van Wilgen, 1986) leading to increased fuel loads in the event of a fire. When fire occurs, the intensity is much greater and may well be sufficient to sterilize the soil, killing the seeds of indigenous plants. High-intensity fires will also lead to water repellency in soils which have been associated with high rates of surface runoff and soil erosion in storms following fires (Scott, 1993; Le Maitre et al, 1996). The hydrological consequences are higher floods and increased siltation of water-courses and reservoirs.

It has been standard forestry practice in South Africa to avoid planting trees in the riparian zones of afforested catchments, to reduce the risk of soil erosion close to the stream channel and to avoid any increase in water use by riparian vegetation. Invaders are no respecters of forestry practice and often spread rapidly into these riparian areas (Dye and Poulter, 1995). It has been demonstrated that removal of infestations of self-established riparian trees can have huge effects on streamflow. Dye and Poulter (ibid) have shown that removal of a strip of self-sown *Pinus patula* and *Acacia mearnsii* along a 500 metre riparian zone at Kalmoesfontein increased streamflow by 120 per cent. Even more dramatic are reports of a stream in Mpumalanga (DWAF, 1996) which, before the clearing of a 500 metre strip of riparian *Eucalyptus grandis*, disappeared within 50 metres of entering the stand. About three weeks after clearing the eucalypts the stream was visible for 200 metres in the stand, and after one month it was running through the stand. It was postulated that it had taken about a month for the stream and rainfall to recharge the water table, restoring a perennial stream from a dry streambed.

The hydrological dangers from invading trees are not just local in their impact. It has been calculated (Le Maitre et al, 1996) that unless curtailed, invaders will eventually reduce the water supply to Cape Town by 30 per cent. It has also been shown that the cost of water from the best dam option is several times more expensive than the cost of water yielded through clearing the invading aliens (DWAF, 1996).

Working for water

> *'Through the clearing of invasive alien plants, the Working for Water Programme is helping us to secure vital water supplies. It typifies the aspirations of the Reconstruction and Development Programme. Already it has brought hope and dignity to thousands of South Africans by creating jobs and business opportunities, and by empowering local communities to care for water and their natural environment.'*
>
> Source: President Nelson Mandela

South Africa's Working for Water Programme is directed at controlling invasive alien species of shrubs and trees for ecological and water conservation purposes, but it is recognized that its greatest challenge is to optimize the opportunities for reconstruction and social development. The well-executed adaptive ecological and hydrological research has provided the base which has given politicians the confidence to initiate such an expensive and innovative IWRM programme, sure in the knowledge that the benefits are going to outweigh the costs. The options of clearing or not clearing the invaders are clearly spelled out in the programme's publicity material, an extract of which is given in Table 5.9.

Table 5.9 *Water resource, ecological and social benefits following the clearing of invasive plants*

No clearing	Twenty-year clearing programme
Water	
Total infestation of catchments in 40–50 years	Far greater long-term water security
Loss of ~3000 million m³ of water per year	Optimal flow in rivers in dry seasons
Cost to replace lost water ~R12000 million	Cost to clear invader plants ~R900 million
Very high price of water and seasonal shortages	More affordable and assured supply of water
Massive soil erosion and siltation of dams	Greatly reduced soil erosion and dam siltation
Greater scouring of rivers and increased flooding	Reduced scouring of rivers and flooding
Ecological	
Extinction of over 1000 plant and animal species	No extinction of plants and animal species
Some rivers, estuaries, wetlands and aquifers will dry up	Enhanced ecological functioning of water systems
Massive impacts of fires on life and property	Greater stability and diversity in ecological systems
Social	
Greater levels of unemployment	4000 direct jobs for 20 years
Increases in crime and non-payment of services	Empowerment, community-building, human dignity
Lost opportunity for social development	Benefits for health, welfare, social stability
Migration of rural communities to urban areas	Migration of urban communities to rural areas
Loss of productive agricultural land	Possible increase in productive agricultural land
Loss of use of wild flowers, thatching grass, herbs	Very profitable harvesting of 'wild' resources

In the opinion of the Department of Water Affairs and Forestry:

> '*by cutting down invading alien plants (such as wattles and pines), we significantly enhance the availability of water. Invasive alien plants grow and spread. If we fail to eradicate them, they will strangle our water supplies. In essence, either we pay now, or we pay more later.*'

Chapter 6

Integrated Water Resources Management

CONCEPTS AND PRINCIPLES

IWRM involves the coordinated planning and management of land, water and other environmental resources for their equitable, efficient and sustainable use. IWRM programmes need to be developed alongside, and not in isolation from economic structural adjustment and other sectoral programmes. For IWRM strategies to be implemented, fragmentation of institutional responsibilities must be reduced.

IWRM objectives encompass the UNCED principles:

- Water has multiple uses and water and land must be managed in an integrated way.
- Water should be managed at the lowest appropriate level.
- Water allocation should take account of the interests of all who are affected.
- Water should be recognized and treated as an economic good.

IWRM strategies seek to ensure:

- A long-term viable economic future for basin dependants (both national and trans-national).
- Equitable access to water resources for basin dependants.
- The application of principles of demand management and appropriate pricing policies to encourage efficient usage of water between the agricultural, industrial and urban supply sectors.
- In the short term, the prevention of further environmental degradation and, in the longer term, the restoration of degraded resources.
- The safeguarding of local cultural heritage and the local ecology as they relate to water management and the maintenance and encouragement of the potential for water related tourism together with linkages between tourism and conservation.

IWRM strategies should recognize that:

- Solutions must focus on underlying causes not merely their symptoms.
- Issues must be approached in an integrated way.
- In general, development of sound resource management and collective responsibility for resources will take place at the sub-regional or village level.

IWRM implementation programmes should:

- Comprise an overall strategy that clearly defines the management objectives, a range of delivery mechanisms that enable these objectives to be achieved, and a monitoring schedule that evaluates programme performance.
- Recognize that the development of water resource management strategies may require research to assess the resource base and, through the use of models and the development of decision support systems, to determine the linkages between water resource development and the impacts on the environment, socio-economics, equity, and ecology.
- Ensure that mechanisms and policies are established that enable long-term support to programmes of environmental recovery.

THE PHILOSOPHY

Increasingly IWRM is regarded as the philosophy or process underlying the management of water resources. Unlike earlier quests for the development of a tightly defined 'water resources master plan', there is increasing recognition that IWRM is more likely to be achieved if it is structured as an incremental, evolving iterative process. Earlier attempts at finding optimal solutions (in the linear programming sense), have been largely abandoned in favour of 'satisficing' or workable solutions involving incremental change without requiring unnecessarily abrupt and disruptive changes. IWRM must accommodate means of obtaining progressive commitments from stakeholders to new developments and initiatives. This new paradigm which is developing is the essence of the blue revolution. Technological, or 'hard', methodologies and solutions cannot ever, on their own, be sufficient. They need to be blended and tempered with 'soft' methods that can deal with the human dimension. They also need to be used with, or be constructed so that they can incorporate socio-economic tools and methodologies which can be used to evaluate and compare different development options

The blue revolution is dependent on these tools and these 'integrating' methodologies and they are being rapidly constructed. They are being recast in both the hard mould which captures all our recent advances in information technology and the physical and environmental sciences, and in the soft mould which contains our new appreciation of the complexity and importance of the human players in unstructured real world systems; systems where tensions and conflicts abound and where the linkages between different interest groups are ill defined. They are being developed so that society's appreciation of land use and water resources in relation to Conservation, Amenity, Recreation and Environment (CARE) issues can be quantified. Most importantly these tools are being designed and constructed to work and mesh together. The success of this design will be of crucial importance to the success and rapid advance of the blue revolution.

SOFT SYSTEMS TOOLS

New developments in the field of operational research and management science have led to a number of new methodologies for dealing with unstructured situations that are intractable by 'hard' methods. These methods include problem structuring methods (PSMs) (Rosenhead, 1996), soft systems methodology (Omerod, 1996) and cognitive mapping. These methods, particularly the soft system methodology, can assist in the development of hydroinformatic and decision support systems by aiding the analysis of organizational activities and information flows and by facilitating the communication between system developers and users (Amezaga and O'Connell, 1998).

PSMs allow ill-defined situations involving humans to be structured to negotiate the way forward. These methods are currently being applied and are in fairly widespread use as management tools in some of the largest, multinational, commercial and marketing organizations. They may well have great relevance in the equally, if not more, complex field of land and water resource management.

Another approach advocated by Simonovic and Bender (1996) is collaborative planning. To facilitate communication between people with different backgrounds, they have developed a collaborative planning-support system (CPSS) which attempts to blend modern computer technologies with modelling and analysis tools in a user-friendly environment. The aim is to enhance the communication between proponents of resource development and those affected. This system can be considered as a module or component of a decision support system, but a module which specifically addresses stakeholder involvement and communication between stakeholders. Simonovic and Bender (ibid) describe the architecture of the module which requires four components: the stakeholders, a list of 'grounded facts', a knowledge base and a potential list of evaluation criteria. The system operates by stakeholders supplying information to the module which describes their individual values or perception of important issues. From this process, the 'grounded facts' are generated and these are used to relate stakeholder values and issues to information useful for the assessment of tradeoffs. Stakeholders can then supply their value information by selecting important features from the list of 'grounded facts'. Each stakeholder can maintain a list of selected facts and, collectively, can duplicate selections. This process indicates areas where there is agreement between stakeholders. The selected lists of 'grounded facts' are then 'translated' through a knowledge base to produce relevant multidisciplinary planning objectives which can then be evaluated using the predefined criteria. Simonovic and Bender (ibid) state that as trust can be a major concern amongst stakeholders the knowledge base, which is usually composed of rules generated by experts who are often proponents of the development, is open to viewing and the source of the rules is made available. Simonovic and Bender (ibid) state that *'Communication is the key in this application. Common understanding and a potentially greater level of consensus is the desired result.'*

Participatory Approaches

Principle Number Two of the Dublin conference report (see Chapter 4) states *'Water development and management should be based on a participatory approach, involving users, planners and policy-makers at all levels'.*

Whilst the participatory approach to water management and development was formalized relatively recently in the Dublin Statement in 1992, participatory approaches had been advocated, and participatory methodologies had been developed, much earlier by social and management scientists in governmental and non-governmental development organizations. Whilst the old mantra of the traditional water engineer may have been 'to meet all reasonable needs', the new mantra of the developmental social scientist is perhaps 'to meet all stakeholders and participate'. Participatory approaches are undeniably a necessary component of the new approach to water management and a key component of the blue revolution. They are not, as zealous social scientists might sometimes suggest, a sufficient condition. There is a danger that participatory approaches are used as the 'stamp of approval' to justify prior conclusions and to deride perhaps perfectly sound solutions and results which have not been achieved through the full participatory process. It is not difficult to stage-manage 'participation' to provide the 'gloss' on decisions and approaches that are being sought by particular pressure groups. Nor is the participatory approach a substitute for the integrated or holistic approach. Interestingly, within the Dublin Statement the holistic approach is enshrined in Principle 1 whilst participation is embodied in Principle 2.

A particular danger is that sectors such as agriculture, forestry and environmental management might adopt the participatory process with religious zeal to foster their own sectoral ambitions, without considering the overall or integrated approach to land use or water resource management.

The World Bank has recently compiled a source book on participatory approaches (see Appendix 2) which details experience in many countries and within different sectors. The source book also provides a valuable glossary of tools which have been developed by social scientists and other development practitioners to encourage and enable stakeholder participation. The source book states that some tools are designed to inspire creative solutions, others are used for investigative or analytic purposes. One tool might be useful for sharing or collecting information, whereas another is an activity for transferring that information into plans or actions.

Indigenous and Vernacular Knowledge

If stakeholder involvement in environmental, land and water management is to be more than tokenism by planning and decision-making authorities, efforts are required to both structure the consultative process and incorporate stakeholder knowledge.

> *'Paradoxically, the planning ethos is open to all science, including "vernacular" science or the "common knowledge of ordinary folk". Adaptive planning, like public response to hazards, requires options and experience; whilst the scientist may set up the valid options, it remains the experience of (and "comfort" with) those options by the public which allow on-line adaptation of the plan to occur.'* (Newson et al, 1998)

The incorporation of indigenous knowledge (IK) in development programmes dealing with natural resource management is discussed by Barr (1998). Although the incorporation of IK is now becoming more common and methodological research on the incorporation of IK within natural resources research is now being developed, Barr warns of the difficulties associated with linking overlapping spheres of knowledge between local people at one end of a spectrum, with the applied and basic sciences at the other, and with social scientists and anthropologists somewhere in the centre. He argues that, in theory, natural resources IK has much to offer in tackling natural resource management problems, but the practical realities of operating this type of interdisciplinary research are far from straightforward. Conflict, or at least disagreement, is recognized as a common feature of interdisciplinary research, especially where the disciplines are 'closely guarded cabals'. Barr suggests that computers may help bridge the paradigmatic differences between researchers and also provide the tools for the incorporation of IK. He cites the work of Walker and Sinclair (1998) where computers have been used to develop cognitive maps of IK in agricultural and forestry systems and work in the social sciences which has led to computer-assisted qualitative data analysis software, known as CAQDAS (Coffey et al,1996).

There may be other dangers in trying to incorporate IK. Barr and Gowing (1998), from experiences of participatory approaches for incorporating IK relating to floodplain management in Bangladesh, warn that the process can be easily 'captured' by locally influential people.

Lack of awareness of, and insensitivity towards, IK may have much graver dangers. Advocacy and the imposition by Western scientists of ineffective agroforestry systems as a claimed means of increasing production in the developing world (see Chapter 2), counter to local traditions, has led to huge wastes in research and development funding and the suspicion and disbelief of local people (and some governments) in development efforts. Newson et al (1998) condemn 'unthinking westernism' for introducing the water closet into some semi-arid countries where it has become a symbol of social rank, although its widespread use would be inherently unsustainable, as another example of development going ahead contrary to local indigenous knowledge and experience.

HARD SYSTEMS TOOLS

Technological tools, which can operate within or link with a 'soft systems' framework, for assisting with the process of IWRM, both for the development of national strategies and for the development and implementation of catchment plans, are now being developed. In earlier chapters it was shown how an understanding of the effects of land use, and land use change, on water resources is one component of the knowledge base that is required to apply principles of IWRM. It was shown how models based on the 'limits' concepts, have been applied for assessing forest impacts on water resources in both wet and dry climates of the world. Wet climate applications, where interception loss predominates, include the uplands of Scotland and the Otago catchments in the South Island of New Zealand. Dry climate applications include a study, reviewed in Chapter 5, of how land use change in Malawi has altered the water balance of Lake Malawi and a regional study, extending the Malawi study, through the use of GIS technology to the Zambezi basin. These land use impact models and other models related to land use and water resources, including the economic, ecological, health and sanitation dimensions, can now be run on common databases through the use of decision support systems (DSS). The use of these systems for advancing land use and integrated water resources management – the blue revolution – is outlined below.

Decision support systems for IWRM

Although a great deal of thought and effort underlies the development of the UNCED principles, and paramount importance is attached to them by governments and UN agencies, much less thought has been given to how these principles can actually be implemented. Agencies and organizations entrusted with the implementation of IWRM are largely at a loss when it comes to knowing how to put in place these procedures. This is a problem not only for the developing world, but also for the developed world where concepts such as stakeholder participation and demand management are still relatively new. Decision support systems have a role to play here in providing mechanisms not only for testing out the impacts of water resource management strategies on stakeholder interests – equity, environmental, ecological and socio-economic impacts – but can also assist water resource managers by providing a focusing framework for defining stakeholder issues and interrelationships.

The greatest strength of these systems is that they provide the means for integrating information from different disciplines: their greatest weakness is that their construction entails a degree of trust, cooperation and professional self confidence from all the disciplines involved, which cannot necessarily be assumed at the outset and which may take considerable time and effort, through the development of a participatory approach, to establish. At the most fundamental level DSSs can be of great value for IWRM through allowing, perhaps for the first time, different data sets from the water resource disciplines (comprising for example surface water, groundwater and water quality) to be displayed together. The hydrologists, hydrogeologists and water chemists who collect these different data sets come from different professional backgrounds and do not necessarily have a tradition or ethos of working closely together and it is not uncommon to find their respective databases residing on different computer systems. Nor is it uncommon to find that these computers are in different buildings with little or no means of transferring or sharing data. With such a poor tradition of integration even amongst the closely related water resource disciplines, the task of integrating and soliciting cooperation amongst the wider disciplines that are involved (environmental science, ecology, socio-economics and health) presents the overriding challenge for DSS development. Ultimately this will also determine if true integrated management of water resources can be achieved.

DSSs bring together different components of hydrologically related information technology within the new science of hydroinformatics (Abbott, 1991). They combine the capabilities of a database, GIS modelling systems and possibly optimization techniques and expert systems, all set within a graphical user interface (GUI).

Two 'milestone' publications (Jamieson, ed, 1996; and O'Callaghan, 1996) describe both the ethos and underlying philosophy of DSSs together with providing outline descriptions of the structure of some of these systems. The capabilities and limitations of regional hydrological models that are used in these systems had earlier been discussed by O'Connell (1995); and the means for integrating these models within DSSs, by Adams (1995). Two of these systems, the WaterWare (Jamieson and Fedra, 1996) and the NELUP (O'Callaghan, 1995) are briefly described below.

Trial applications of the systems were carried out on the upland Tyne river basin in the north of England and the lowland Cam river basin in the east of England. Subsequent development of the WaterWare and NELUP DSSs has led to applications in the Mediterranean area to investigate Mediterranean desertification and land use under the EU-funded MEDALUS programme. This programme is focused on the Agri river basin in Italy and integrates both physical and socio-economic models.

The WaterWare DSS for Integrated River Basin Planning

A decision support system designed to meet the needs of the European water industry was developed using European Union funding under the EUREKA EU 487 programme. Three universities, a research institute and two commercial companies were involved in the development.

WaterWare Objectives
The system was designed to be capable of addressing a wide range of issues including:

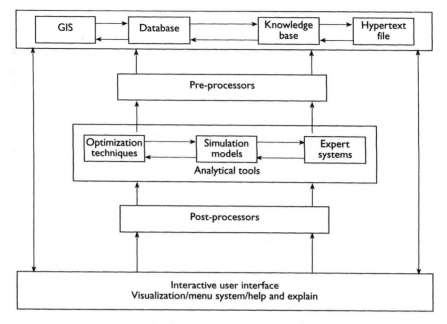

Figure 6.1 *WaterWare System Architecture*

1 Determining the limits of sustainable development.
2 Evaluating the impact of new environmental legislation.
3 Deciding what, where and when new resources should be developed.
4 Assessing the environmental impact of water related development.
5 Formulating strategies for river and groundwater pollution control schemes.

The system was developed to run under the UNIX operating system, making extensive use of colour graphics, hypertext links and is largely icon driven for ease of use; the system architecture is shown in Figure 6.1.

WaterWare Database and GIS

The database is able to hold data of a mixture of types, all of which are geo-referenced to a GIS. These include static or object oriented data that are fixed in time and could include specifications and details and photographs of, for example, a treatment works or flow measurement station. Slowly changing data such as the treatment processes used or works configuration data are also accommodated, as are rapidly changing time series data such as daily flows or rainfall. Time series data can be displayed graphically in a number of ways and routines are available for statistical analysis of the data including calculation of the mean, standard deviation, cross correlation coefficients and double mass curves.

The GIS contains all the spatial data on land use, elevation, geology, soil type, river and road networks in the appropriate raster or vector format. Options are available for specifying the sequence in which spatial files are overlaid and edited and a zoom facility is also incorporated.

WaterWare Modelling Components

The standard WaterWare system has modelling components for groundwater and surface water pollution control, irrigation and domestic water demand forecasting, hydrological processes and water resource planning. For groundwater pollution control, a two-dimensional finite-difference model of flow and contaminant transport is incorporated to predict the movement of pollutants.

The decision support module comprises a three-stage process: an expert system to locate possible sites for scavenger boreholes; followed by an artificial neural network for assessing the performance of different combinations of boreholes; and finally a generic algorithm for selecting the most cost-effective solution for reducing the pollution to a prescribed residual level.

For surface-water pollution control and for river restoration schemes, the DSS contains a one-dimensional stochastic river water quality model.

To achieve the lowest-cost solution for achieving a prescribed river water quality standard, a waste-load allocation module is configured to operate with heuristic search and linear programming algorithms to select the appropriate technologies for each effluent discharge. Alternatively the module can be run to identify the most effective allocation for improving river quality for a fixed budget.

For water demand forecasting, either for irrigation or domestic supply purposes, the DSS incorporates an expert system shell. The FAO's CROPWAT model has been rewritten as a rule-based inference module within this shell to calculate irrigation demand requirements. All the data requirements are contained within the database including GIS, topography, land-use, soil-type, rainfall and potential evaporation. The module, given fertilizer and water supply costs, is also able to calculate the economic returns from different combinations of crop and irrigation system (for example surface, spray and drip).

The hydrological processes component contains a spatially aggregated daily rainfall/runoff model which is able to generate simulated flow records for ungauged tributaries.

For more detailed modelling of the rainfall/runoff processes, the SHETRAN modelling system (Ewen et al, 1999) can also be used.

For water resource planning, a generic model was developed, operating on a daily time step, to simulate the balance between supply and demand and to indicate the frequency and extent of shortfalls. The model can be used to assess the water resource system and can be used with a screening model to minimize costs, both monetary and environmental, for the development of regional and national water plans.

The WaterWare system was first applied to the Thames basin in southern England as a case study. Other applications include the Rio Lerma/Lake Chapala master plan in Mexico and water resource planning in the Occupied Palestinian Territories.

The NELUP decision support system

The joint Natural Environment Research and Economic and Social Research Councils' land use programme, under which the NELUP DSS was developed, was an innovative programme designed to investigate the interactions between land use, water resources, economics and ecology. Whilst the WaterWare system is oriented more specifically towards the water industry for the planning and operation of water developments within river basins, the NELUP DSS is aimed at a higher level of land and water resource management which is more closely aligned with the wider objectives of IWRM.

NELUP Design Objectives

This decision support system was designed to:

1   Integrate models covering economics, ecology and hydrology that describe the changes in the spatial pattern of land use and the impacts of these changes.
2   Integrate nationally available data sets which describe the biophysical and economic conditions within a river basin in a database.
3   Create an interactive, user-friendly interface to the database and models to allow exploration of future land use scenarios.

NELUP Database and GIS

To take into account both water resource, socio-economic and ecological issues related to land use requires extensive and compatible databases. In the NELUP DSS, spatial data are held within a raster GIS (GRASS) and non-spatial data are held within a relational database management system (ORACLE). The data sets include agricultural data, parish census data, farm business survey data, national vegetation data, digital river networks and river gauging records, but central and common to all the modelling routines are the four datasets comprising:

1   Digital elevation, derived from digitized contours of Ordnance Survey 1:50,000 maps (used for the definition of catchment boundaries and for the definition of land capability maps for the catchment level and farm level economics and for ecological models).
2   Land cover classified into 25 land cover types derived from Landsat imagery.
3   Soils data obtained from soil association maps and soil series parameter data which were provided by the Soil Survey of England and Wales.
4   Meteorological data obtained from the Meteorological Office.

The Models

Three types of model are incorporated in the NELUP DSS: agro-economic, ecological and hydrological. These models can be used to investigate the characteristics of a region under its present land use and to evaluate how these characteristics will respond to specific land use changes. A land use change can be specified explicitly by direct intervention of the user of the DSS or implicitly as the result of economic policy shifts which encourages land owners to make changes in land use and land management.

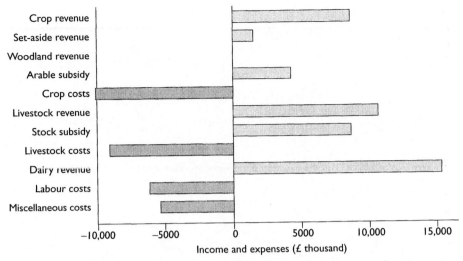

Figure 6.2 *Predicted income and expenses for the Tyne river basin*

The agro-economic, linear programming models operate at two spatial scales, at the catchment scale and at the farm-level scale. The catchment-level model predicts the aggregated behaviour of a region by treating the area as a macro farm. It is capable of taking into account the wide variation in agricultural activity that can exist across a region. Predicted land use patterns can be disaggregated to a finer spatial resolution by categorizing the region according to its agricultural production potential. Given inputs of values for crop prices, agricultural subsidies, and costs of materials, the model will calculate changes in land use patterns and income across the region, as for example in the Tyne river basin (Figure 6.2).

The farm-level agro-economic model is similar in operation to the catchment-level model but it has the capability of investigating the potential response of a particular type of farm to changes in agricultural policy. Land use change is modelled at the local rather than the regional scale.

Three types of ecological model (associative, vegetation environment management, and bird and mammal) are used to predict the distribution of various species within the landscape and how land use and environmental change will impact on these species distributions.

An associative model is used to predict the distribution of both plants and invertebrates by linking species distribution to their known occurrence in particular land cover types, through an ecological hierarchy comprising land cover, community and species. An example concerning perennial rye-grass in the Tyne river basin is shown in Figure 6.3.

Variations in the distribution of species are predicted directly from a knowledge of the change in land cover.

A vegetation environment management model is used to indicate how changes in agricultural management, such as grazing pressure and fertilizer applications, will change species distributions. This model can also be linked to the farm-level economics model to predict the eventual impact of economic policy shifts on vegetation distribution and diversity.

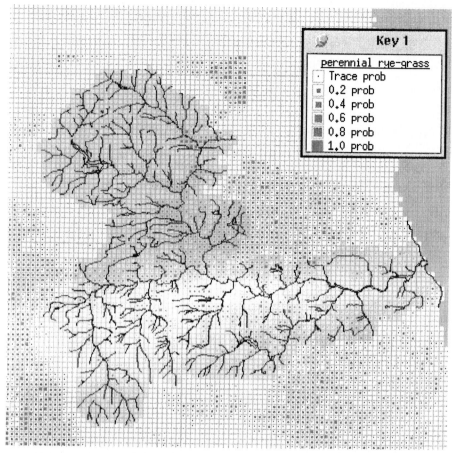

Figure 6.3 *Probability of perennial rye-grass in the Tyne river basin*

A bird and mammal ecological model relates species distributions not just to individual land covers but, because these species are highly mobile, to the combination of land covers that they inhabit.

Two hydrological modelling systems are available in NELUP, the SHETRAN-distributed, physically-based system, which is also available in the WaterWare DSS and the NUARNO system which uses a spatially aggregated approach. In some respects the scales over which they would normally be applied are analogous to the scales used for the agro-economic models. NUARNO provides an overview of the catchment hydrology whilst SHETRAN has the capability for analysing smaller scales in much greater detail.

Hydrological impacts are calculated by running the models with parameters relating to the existing land use to determine the existing hydrological regime (see example from the Tyne in Figure 6.4) and then running the models again with the new land use scenario, determined either by the agro-economic or the ecological models.

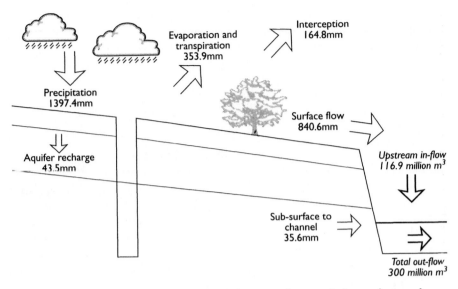

Figure 6.4 *Predicted components of the annual water balance for a sub-catchment of the Tyne*

SOCIO-ECONOMIC METHODOLOGIES – EVALUATING DEVELOPMENT OPTIONS

The processes which determine how land use impacts on water resources are now becoming understood (see Chapters 1 and 2) and methods are now available for quantifying some of these impacts, particularly as they relate to forests (see Chapter 3). Methods have also been developed to attribute an economic value to both water in it various uses and to land use in both its primary and secondary uses (see Chapter 4).

Management of land use and water resources requires this knowledge, and better knowledge, of the interactions between land use and water resources, but it also requires methods through which this knowledge can be used so that the relative benefits of potential water resource developments or land use changes can be assessed. These methods are required in local, regional and international contexts to assist the production of plans, guidelines, strategies and national and international policies and legislation. Increasingly the 'secondary' uses of land and water, which can be grouped under Conservation, Amenity, Recreation and Environment (CARE), are assuming greater importance and these methods must be able to take these uses into account.

Socio-economic methodologies provide the necessary framework for evaluating and comparing different development options. These methodologies are being developed for specific land uses such as forests (Benson and Willis, 1991; Willis and Benson, 1989) and, more recently for agricultural lands (Harvey and Willis, 1997). Socio-economic methodologies are also being developed (ERM and University of Newcastle upon Tyne, 1997) to assist the economic appraisal of the environmental costs and benefits of possible solutions for ameliorating low flows in rivers.

The concept of IWRM is relatively new to land use and water resources managers. These methodologies are even newer. It is important that these embryo evaluation frameworks, arguably crucial to the blue revolution, are fostered. They will need to be robust and defensible and they need to be understood by their practitioners so that they are not misapplied.

Use and Abuse of Socio-Economic Methods

The English and Welsh Environment Agency (EA), the regulatory body, has a statutory duty to consider costs and benefits in exercising its powers (see Chapter 4). In collaboration with the Department of Environment, Transport and the Regions (DETR), the water companies regulator (Ofwat) and the water companies, it has developed a manual on approved methods of assessing the benefits of water schemes, known as the benefits manual (Foundation for Water Research, 1996).

The value, use and abuse of these socio-economic methodologies is well illustrated by their disposition in the conflict situation and inquiry which arose as a result of the application by Thames Water to abstract more groundwater at Axford in the environs of the River Kennet, a chalk stream in Wiltshire in the south of England. The stream is regarded as of high environmental quality, being both in an Area of Outstanding Natural Beauty and an SSSI; it is also valued by anglers as a trout river.

Thames Water applied to make permanent a temporary variation to its abstraction licence which had been in force for 15 years. The base licence entitled it to take up to 13.1 megalitres per day ($Ml.d^{-1}$) and the variation entitled it to take a further 7.4 $Ml.d^{-1}$ subject to a minimum flow of 61.4 $Ml.d^{-1}$ being achieved in the river. The EA response was to impose a progressive increase in the minimum flow requirement which would have the effect of both cutting permitted abstractions at periods of low flows and reducing the base licensed volume.

Thames Water appealed on the grounds that the abstractions had a marginal effect on river flows and that recent changes in the river's ecology were due to drought, dredging of the river channel and management of water levels by anglers.

The appeal forced the EA into preparing a cost-benefit assessment, following the guidance of the benefits manual. The recommended cost-benefit analysis considers the public's 'willingness to pay' (WTP) for both 'use' benefits accrued from those using the river directly through activities such as angling and recreation, and 'non-use' benefits accrued from its aesthetic, cultural, conservation or legacy values, which can be realized without the need to visit the river. In retrospect, perhaps unwisely, the EA decided that a full WTP assessment, involving extensive public opinion surveys, would have been too costly and time consuming and it opted for a 'benefit transfers' approach which made use of previously collected WTP figures on other low-flow rivers such as the Darent in Kent. Using transferred WTP figures, the EA calculated the use value of the Kennet at £67,000 per year. Using this figure, with a discounting period of 30 years, the additional benefits of the river if Thames Water's abstractions were reduced, was calculated at a net present 'use' value of £0.4 million. The data from the Darent study showed that people up to 60 kilometres distant would pay an average of 32 pence per household per year per river to counter low flows. Again transferring these data to the Kennet, the EA multiplied these figures by the number of domestic water connections in Thames Water's supply area (three million) and, again discounting over 30 years, arrived at a net present 'non-use' value of £13.2 million. Thus the total calculated value of reducing abstractions was £13.6 million.

The 'non-use' values for the river were therefore set by the EA at a figure about 30 times the 'use' values. This high value comes about through the contentious use of the figure of three million water connections, as being appropriate for WTP calculations. Thames Water argued that its use was nonsensical as, had a smaller company with fewer connections been involved, the valuation would have been much less. They argued that the river was mainly a local issue and claimed only 100,000 people were affected. In the benefits manual it is recommended that where the outcome of a cost-benefits analysis may be dependent on the 'non-use' values, new fieldwork may be justified to improve confidence in the 'non-use' figures. The question of what population to consider in calculating 'non-use' values appears to be a fundamental problem in benefit assessment (END, 1998).

At the subsequent public inquiry the Environment Inspector overruled the EA and found in favour of Thames Water on this crucial question of how to calculate 'non-use' WTP. The inspector assigned the 'non-use' value at only £0.3 million. As Thames Water had calculated that the cost of replacing the abstractions with water from other sources would be £6.2 million there was a substantial economic margin in favour of continuing with the groundwater abstractions. The Environment Secretary, in February 1998, endorsed the Inspector's findings and upheld Thames Water's appeal. The final outcome was a partial compromise as the inspector conceded that the Kennet's ecology was being harmed by low flows and ruled that the minimum river low flow should be raised from 61.4 to 90Ml.d^{-1}. Nevertheless, this was less than the 104Ml.d^{-1} sought by the EA.

The Axford inquiry highlights the present dilemma in the use of socio-economic methods for evaluating development options. Perhaps most seriously, misuse of these methods by the EA has brought their application into disrepute. The Darent study was not designed to yield transferable estimates. Even if it had been, considerable problems would remain in the calibration and aggregation over the relevant populations associated with the Kennet.

At present, methods for calculating transferable benefit estimates are in their infancy although the study of low-flow rivers in the south-west of England (ERM and University of Newcastle upon Tyne, 1997) went some way towards generating transferable estimates for low-flow improvements. The production of robust, defensible and transferable methodologies for benefit evaluation to assist in the choice of development options remains a high priority for the blue revolution. Without these methodologies, a rational framework for setting development options in a form where production benefits can be compared against economic, ecological, equity and socio-economic considerations is lost to us.

IWRM IMPLEMENTATION AND FUTURE DEVELOPMENTS

Although the decision support systems that have been developed have not so far been applied worldwide, their structure is sufficiently general for their usage in both the developed and developing world. These systems are also currently being developed to run on PC as well as UNIX platforms (on which they were originally developed) which will make them more suitable in developing-world environments. Developments are also planned to combine the components of these two systems with the inclusion of 'limits' concept land-use evaporation models. This would allow the construction of 'bespoke' systems, tailored to a particular country's needs. With economic, environment, ecological and water resource components built into these models, which truly reflect the particular developments within a basin, the catchment planner or agency responsible for planning developments will have the tools at hand to assist with the implementation of the UNCED principles.

The ideals of IWRM, the basis of the blue revolution, are now well developed, recognized and accepted and technological advances that have led to the development of decision support systems give us some of the tools for implementation – yet progress is slow. Water affects so many aspects of development yet water specialists tend to be specialist in just a particular aspect of hydrology. The development and application of 'soft system methodologies' to go alongside the technological advances of the blue revolution hold out great hopes for the future.

The biggest challenge for IWRM and the blue revolution is to ensure not only user involvement, but also that the professionals who can influence decisions in land use and water resources, including water resource and environmental managers, economists, socio-economists, ecologists and health and sanitation workers and engineers, are all prepared to contribute and collaborate. Only by doing so, can sustainable and robust solutions be achieved for managing our environment and water systems for the future.

Appendix 1

IWRM Contact and Linking Organizations

- *The African Water Page.* This is a website dedicated to the water sector in Africa. Issues addressed include water policy, water resource management, water supply and environmental sanitation, water conservation and demand management, and a variety of other issues. The stated objective of the site is information dissemination on water issues in Africa and to exchange views and ideas on water on the continent.

  The website address is http://wn.apc.org/afwater/index.htm

- *The American Heritage Rivers Initiative.* The American Heritage Rivers initiative was set up to help communities revitalize their rivers and riverbanks, through conserving historic buildings and natural habitats with a focus on history and heritage.

  The website address is http://www.epa.gov/rivers/

- *California Resources Agency.* The California Resources Agency is responsible for the conservation, enhancement and management of California's natural and cultural resources, including land, water, wildlife, parks, minerals and historic sites. The agency is composed of departments, boards, conservancies, commissions and programmes. The agency has developed an information system, CERES, to improve environmental analysis and planning by integrating natural and cultural resource information from multiple contributors and by making it available and useful to a wide variety of users.

  The website address is http://ceres.ca.gov/cra/ http://ceres.ca.gov/cra/

- *CGIAR.* The Consultative Group on International Agriculture Research (CGIAR) is a consortium jointly supporting a system of 16 international agricultural research centres. The mission of the CGIAR is to contribute, through research, to the promotion of sustainable agriculture for food security in the developing countries. Two centres of particular interest to water related issues are, the International Irrigation Management Institute (IIMI) in Sri Lanka, and the International Food Policy Research Institute (IFPRI) in Washington, USA.

  The website address is http://www.cgiar.org/

- *CLUWRR.* The Centre for Land Use and Water Resources Research (CLUWRR) spans three faculties of the University of Newcastle upon Tyne: Engineering, Agriculture and Biological Sciences, and Social and Environmental Sciences. It carries out multidisciplinary research into land and water issues both in the UK and worldwide. It was the principal agent for the development of the NELUP decision support system which takes an integrated approach to land and water management through the operation of hydrological, economic and ecological models within a common spatial data base.

  The website address is http://www.cluwrr.ncl.ac.uk/

- *CSIRO Water Resources Division.* The Water Resources Division is a division of Australia's major government-supported national research organization, the Commonwealth Scientific and Industrial Research Organization (CSIRO) with centres in Canberra, Albury, Griffith, Adelaide and Perth.

  The website address is http://www.csiro.au/

- *DFID.* The Department for International Development (DFID) of the British Government has, as it overall objective, the elimination of poverty in poorer countries. It is committed to the development of efficient and well-regulated markets, access of poor people to land, resources and markets, safe drinking water and food security, the sustainable management of physical and natural resources, efficient use of productive capacity and the protection of the global environment.

  The website address is http://www.dfid.gov.uk/

- *DWAF.* The Department of Water Affairs and Forestry (DWAF) in South Africa, is a world leader in applied water research and water management models.

  The website address is http://www-dwaf.pwv.gov.za/idwaf/web-pages/minister/minister.htm

- *EA.* The English and Welsh Environment Agency (EA) is the largest environmental body in Europe. It is committed to achieving the objectives of integrated catchment management and development and is a key player in furthering the Government's sustainable development policies. It also has the responsibility for regulating water resources and for authorizing abstractions and effluent discharges.

  The website address is http://www.environment-agency.gov.uk/

- *Environment and Heritage.* The goal of the Environment and Heritage Department of the government of Australia is to develop, in the national interest, a proper recognition of environmental, social and related economic values in government decision-making and activities.

  The website address is http://www.environment.gov.au/

- *EPA.* The USA Environmental Protection Agency (EPA) is an independent agency created by Congress to protect public health and to safeguard and improve the natural environment – air, water, and land. The EPA manages the American Heritage Rivers Initiative (see above).

  The website address is http://www.epa.gov/

- *FAO.* The Food and Agriculture Organization of the United Nations (FAO) is much involved in land and water development worldwide, often with an inclination towards the agricultural use of water, but increasingly also on management issues.

  The website address is http://www.fao.org/

- *GARNET.* The Global Applied Research Network (GARNET) is maintained by the Water Engineering and Development Centre (WEDC) (see below) as a mechanism for information exchange in the water supply and sanitation sector using low-cost, informal networks of researchers, practitioners and funders of research.

  The website address is http://info.lut.ac.uk/departments/cv/wedc/garnet/grntover.html

- *Great Lakes Commission.* The aim of the Great Lakes Commission is to promote the orderly, integrated and comprehensive development, use and conservation of the water resources of the Great Lakes Basin in the USA.

  The website address is http://www.glc.org/

- *GWP.* The Global Water Partnership (GWP) is an international network open to all involved in water resources management, including governments of developing as well as developed countries, UN agencies, multilateral banks, professional associations, research organizations, the private sector and NGOs. The objective of the GWP is to translate the Dublin-Rio principles into practice.

  The website address is http://www.gwp.sida.se/

- *HR.* Hydraulics Research (HR) Wallingford is an independent organization specializing in civil engineering and environmental hydraulics, and the problems of water management.

  The website address is http://www.hrwallingford.co.uk/

- *IAHS.* The International Association for Hydrological Science is an international nongovernmental organization which deals with hydrology and water resources. It was established in 1922, incorporating the International Commission of Glaciers which had been set up in 1894, with the aim of bringing together hydrologists from all countries to promote the hydrological sciences.

  The website address is http://www.wlu.ca/~wwwiahs/index.html#IAHS

- *IAWQ.* The International Association on Water Quality (IAWQ) is a professional membership association dedicated to the advancement of the science and practice of water pollution control and water quality management worldwide.

  The website address is http://www.iawq.org.uk/

- *ICID.* The International Commission on Irrigation and Drainage (ICID) is an international scientific and technical NGO, dedicated to improving water and land management for the enhancement of the worldwide supply of food and fibre for all people. The objectives of ICID are to stimulate and promote the development and application of the arts, sciences and techniques of engineering, agriculture, economics, ecology and social sciences in managing water and land resources for irrigation, drainage, flood control and river training and for research in a more comprehensive manner adopting up-to-date techniques.

  The website address is http://fserv.wiz.uni-kassel.de/kww/irrisoft/coop/icid.html

- *ICOLD.* The International Commission on Large Dams is a non-governmental International Organization which provides a forum for the exchange of knowledge and experience in dam engineering. ICOLD's aims are to ensure that dams are built safely, efficiently, economically, and without detrimental effects on the environment.

  The website address is http://www.icold-cigb.org

- *IH.* The Institute of Hydrology (IH) employs 180 scientists to investigate a combination of fundamental science and applied science to solve practical problems for government departments, international agencies and independent organizations.

  The website address is http://www.nwl.ac.uk/ih/

- *IPTRID.* The International Programme on Technology Research in Irrigation and Drainage (IPTRID) has been set up to enhance irrigation and drainage technology in developing countries. The Programme was initiated in 1991 in response to calls within the profession for a new initiative to enhance and expand research in developing countries. IPTRID's sponsors are the World Bank, UNDP, ICID, bilateral aid ministries and development foundations.

  The website address is http://www.ilri.nl/iptrid-n.html

- *IRC.* The International Water and Sanitation Centre (IRC) focuses on affordable technologies to provide clean water and adequate sanitation to poor people around the world. Working with partners in developing countries, IRC aims to strengthen local capacities by sharing information and experience and developing resource centres.

  The website address is http://www.oneworld.org/ircwater/

- *IUCN.* The World Conservation Union was established in 1948 as the International Union for the Protection of Nature (IUPN) and became the International Union for the Conservation of Nature (IUCN) in 1956. Today it is a union of governments, government agencies, and NGOs working at the field and policy levels, together with scientists and experts, to protect nature.

    The website address is http://www.iucn.org/

- *IWSA.* The International Water Supply Association (IWSA) is an international organisation concerned with the provision of water supply and wastewater management to homes, industry and agriculture. Membership is drawn from a hundred countries across five continents.

    The website address is http://www.iwsa.org.uk/

- *ODI.* The Overseas Development Institute (ODI) is an independent non-governmental centre for the study of development and humanitarian issues and a forum for discussion of the problems facing developing countries.

    The website address is http://www.oneworld.org/odi/

- *SEI.* The Stockholm Environment Institute (SEI) was established in 1989 as an independent foundation for the purpose of carrying out global and regional environmental research. The Institute is active in promoting international initiatives on environment and development issues.

    The website address is http://www.sei.se/

- *SustainAbility.* 'SustainAbility' is a strategic management consultancy and think-tank dedicated to promoting the business case for sustainable development – satisfying the 'triple-bottom line'.

    The website address is http://www.sustainability.co.uk/sustainability.htm

- *TECCONILE.* TECCONILE was set up to assist participating countries in the development, conservation and use of the Nile basin water resources in an integrated and sustainable manner.

    The website address is http://www.tecconile.org/t1.htm

- *UNDP.* The United Nations Development Programme (UNDP) is involved in water-related development activities worldwide. Capacity building, for example in the field of water, is a major focal area.

    The website address is http://www.undp.org/

- *UNEP.* The United Nations Environment Programme (UNEP) was established as the environmental 'conscience' of the UN system. Its mission is to provide leadership and encourage partnerships in caring for the environment by inspiring, informing and enabling nations and people to improve their quality of life without compromising that of future generations.

    The website address is http://www.unep.ch/

- *VKI*. The Water Quality Institute (VKI) in Denmark, is a large multidisciplinary research and consultancy organization focusing on water quality and water quantity issues.

  The website address is http://www.vki.dk/

- *WEDC*. The Water Engineering and Development Centre (WEDC) at Loughborough University is a training and consultancy organization directed towards providing appropriate technologies for clean water and sanitation in the developing world.

  The website address is http://info.lboro.ac.uk/departments/cv/wed/index.html

- *WB*. The World Bank (WB) has an informative and useful web site (http://www.worldbank.org/). Regarding water-related issues, much information can be found on water markets and pricing principles, water management models, case studies, and recent publications and news. A search engine is available.

- *WBCSD*. The World Business Council for Sustainable Development (WBCSD) aims to develop closer cooperation between business, government and all other organizations concerned with the environment and sustainable development.

  The website address is http://www.wbcsd.ch/aboutus.htm

- *WRC(SA)*. The Water Research Commission (WRC) in South Africa does not undertake any research on its own, but it provides funding for research undertaken by institutions like universities, government departments, and industry. Furthermore WRC monitors, directs and coordinates water research in South Africa and disseminates the findings. It is very well placed to assist in linking water research in South Africa with international initiatives. A comprehensive list of linkages on water in southern Africa is provided.

  The website address is http://www.wrc.org.za/

- *WRSRL*. The Water Resource Systems Research Laboratory (WRSRL) is part of the Department of Civil Engineering at the University of Newcastle upon Tyne. It carries out water-related research and training across the full fundamental spectrum of strategic and applied research. The department was awarded the highest possible national research rating in 1996. It has developed a suite of advanced modelling systems for generating rainfall fields, predicting fluxes of water, sediments and contaminants through river basins and modelling groundwater movement and contaminant migration. The WRSRL is also a leader in the development of decision support systems and has been a partner in the development of the WATERWARE and NELUP systems.

  The website address is http://www.ncl.ac.uk/wrgi/wrsrl/

- *WSSCC*. The Water Supply and Sanitation Collaborative Council is a group of professionals from developing countries, external support agencies (ESAs), and non governmental and research organisations all working in the water, sanitation and waste management sector. Its mission is to 'enhance collaboration among developing countries and ESAs so as to accelerate the achievement of sustainable water, sanitation, and waste management services to all people, with special attention to the poor'.
  The website address is http://oneworld.org/ircwater/council.

- *WWC*. The World Water Council (WWC) is an international water policy think-tank with headquarters in Marseilles, France. The Council seeks to develop and implement strategies and policies that will achieve sustainable water resources management for communities worldwide.
  The website address is http://www.worldwatercouncil.org/

Appendix 2

Glossary of Participatory Tools

This glossary is taken from the World Bank sourcebook on participatory approaches.

- *Access to resources.* A series of participatory exercises that allows development practitioners to collect information and raises awareness among beneficiaries about the ways in which access to resources varies according to gender and other important social variables. This user-friendly tool draws on the everyday experience of participants and is useful to men, women, trainers, project staff and field-workers.

- *Analysis of tasks.* A gender analysis tool that raises community awareness about the distribution of domestic, market and community activities according to gender and familiarizes planners with the degree of role flexibility that is associated with different tasks. Such information and awareness are necessary to prepare and execute development interventions that will benefit both men and women.

- *Focus-group meetings.* Relatively low-cost, semi-structured, small-group consultations (four to twelve participants plus a facilitator) used to explore people's attitudes, feelings, or preferences, and to build consensus. Focus-group work is a compromise between participant observation, which is less controlled, lengthier and more in-depth, and pre-set interviews, which are not likely to attend to participants' own concerns.

- *Force field analysis.* A tool similar to 'Story With a Gap', force field analysis engages people in defining and classifying goals and making sustainable plans by working on thorough 'before and after' scenarios. Participants review the causes of problematic situations, consider the factors that influence the situations, think about solutions, and create alternative plans to achieve solutions. The tools are based on diagrams or pictures, which minimize language and literacy differences and encourage creative thinking.

- *Health-seeking behaviour.* A culturally sensitive tool for generation of data about health care and health-related activities. It produces qualitative data about the reasons behind certain practices as well as quantifiable information about beliefs and practices. This visual tool uses pictures to minimize language and literacy differences.

- *Logical Framework or LogFRAME.* A matrix that illustrates a summary of project design, emphasizing the results that are expected when a project is successfully completed. These results or outputs are presented in terms of objectively verifiable indicators. The logical framework approach to project planning, developed under that name by the USA Agency for International Development, has been adapted for use in participatory methods such as ZOPP and TeamUP. ZOPP, from the German term "Zielorientierte Projektplanung", is a project planning and management method that encourages participatory planning and analysis throughout the project cycle. The TeamUP process assists stakeholders in planning and decision making and encourages stakeholders to collaborate as an effective working group.

- *Mapping.* A generic term for gathering baseline data on a variety of indicators in pictorial form. This is an excellent starting point for participatory work because it gets people involved in creating a visual output that can be used immediately to bridge speech gaps and to generate lively discussion. Maps are useful as verification of secondary source information, as training and awareness raising tools, for comparison and for monitoring of change. Common types of maps include health maps, institutional maps (Venn diagrams) and resource maps.

- *Needs assessment.* A tool that draws out information about people's varied needs, raises participants' awareness of related issues, and provides a framework for prioritizing needs. This tool is an integral part of gender analysis, in developing an understanding of the particular needs of both men and women and carrying out comparative analysis.

- *Participant observation.* is a fieldwork technique used by anthropologists and sociologists to collect qualitative and quantitative data that leads to an in-depth understanding of people's practices, motivations, and attitudes. Participant observation entails investigating the project background, studying the general characteristics of a beneficiary population, and living for an extended period among beneficiaries, during which interviews, observations, and analyses are recorded and discussed.

- *Pocket charts.* Pocket charts are investigative tools that use pictures as stimuli to encourage people to assess and analyse a given situation. Through a 'voting' process, participants use the chart to draw attention to the complex elements of a development issue in an uncomplicated way. A major advantage of this tool is that it can be put together with whatever local materials are available.

- *Preference ranking.* This is also called direct matrix ranking, an exercise in which people identify what they do and do not value about a class of objects (for example, tree species or cooking fuels). Ranking allows participants to understand the reasons for local preferences and to see how values differ among local groups. Understanding preferences is critical for choosing appropriate and effective interventions.

- *Role playing.* Role playing enables people to creatively remove themselves from their usual roles and perspectives to allow them to understand choices and decisions made by other people with other responsibilities. Ranging from a simple story with only a few characters to an elaborate street-theatre production, this tool can be used to acclimatize a research team to a project setting, train trainers, and encourage community discussions about a particular development intervention.

- *Seasonal diagrams or seasonal calendars.* These calendars show the major changes that affect a household, community, or region within a year, such as those associated with climate, crops, labour availability and demand, livestock and prices. Such diagrams highlight the times of constraints and opportunity, which can be critical information for planning and implementation.

- *Secondary data review.* Also called desk review, this is an inexpensive, initial inquiry that provides necessary background information. Sources include academic theses and dissertations, annual reports, archival materials, census data, life histories, maps and project documents.

- *Semistructured interviews.* These are also called conversational interviews and are partially structured by a flexible interview guide with a limited number of preset questions. This kind of guide ensures that the interview remains focused on the development issue at hand while allowing enough conversation so that participants can introduce and discuss relevant topics. These tools are a deliberate departure from survey-type interviews with lengthy, predetermined questionnaires.

- *Sociocultural profiles.* Sociocultural profiles are detailed descriptions of the social and cultural dimensions that in combination with technical, economic and environmental dimensions serve as a basis for design and preparation of policy and project work. Profiles include data about the type of communities, demographic characteristics, economy and livelihood, land tenure and natural resource control, social organization, factors affecting access to power and resources, conflict resolution mechanisms, and values and perceptions. Together with a participation plan, the sociocultural profile helps ensure that proposed projects and policies are culturally and socially appropriate and potentially sustainable.

- *Surveys.* Surveys comprise sequences of focused, predetermined questions in a fixed order, often with predetermined, limited options for responses. Surveys can add value when they are used to identify development problems or objectives, narrow the focus or clarify the objectives of a project or policy, plan strategies for implementation, and monitor or evaluate participation. Among the survey instruments used in World Bank work are firm surveys, sentinel community surveillance, contingent valuation, and priority surveys.

- *Tree diagrams.* Tree diagrams are multipurpose, visual tools for narrowing and prioritizing problems, objectives, or decisions. Information is organized into a tree-like diagram that includes information on the main issue, relevant factors, and influences and outcomes of these factors. Tree diagrams are used to guide design and evaluation systems, to uncover and analyse the underlying causes of a particular problem, or to rank and measure objectives in relation to one another.

- *Village meetings.* Village meetings have many uses in participatory development, including information sharing and group consultation, consensus building, prioritization and sequencing of interventions, and collaborative monitoring and evaluation. When multiple tools such as resource mapping, ranking, and focus groups have been used, village meetings are important venues for launching activities, evaluating progress, and gaining feedback on analysis.

- *Wealth ranking.* Wealth ranking is also known as wellbeing ranking or vulnerability analysis. It is a technique which allows for the rapid collection and analysis of specific data on social stratification at the community level. This visual tool minimizes literacy and language differences of participants. Participants are asked to consider factors such as ownership of, or rights to, productive assets; the lifecycle stage of members of the productive unit; relationship of the productive unit to locally powerful people; availability of labour; and indebtedness.

- *Workshops.* Workshops are structured group meetings at which a variety of key stakeholder groups, whose activities or influence affect a development issue or project, share knowledge and work toward a common vision. With the help of a workshop facilitator, participants undertake a series of activities designed to help them progress toward the development objective (such as consensus building, information sharing, prioritization of objectives and team building). In project as well as policy work, from pre-planning to evaluation stages, stakeholder workshops are used to initiate, establish, and sustain collaboration.

References

Abbott, MB (1991) *Hydroinformatics, information technology and the aquatic environment*, Avebury Technical, Aldershot, England

Abbott, MB, Bathurst, JC, Cunge, JA, O'Connell, PE and Rasmussen, J (1986) 'An introduction to the European Hydrological System – System Hydrologique Europeen', 'SHE', 1: History and philosophy of a physically-based, distributed modelling system, *J Hydrol*, 87, pp45–59

Adams, R (1995) 'The integration of a physically based hydrological model within a decision support system to modelling the hydrological impacts of land use change', *Scenario Studies for the Rural Environment*, pp209–214, Kluwer Academic Publishers, The Netherlands

Adams, B, Grimble, R, Shearer, TR, Kitching, R, Calow, R, Chen Dong Jie, Cui Xiao Dong and Yu Zhong Ming (1994) *Aquifer overexploitation in the Hangu region of Tianjin, People's Republic of China*, British Geological Survey, Nottingham

Allan, JA (1992) Fortunately there are substitutes for water: otherwise our hydropolitical futures would be impossible, Paper 2, in *Proceedings of the Conference on Priorities for Water Resources Allocation and Management*. Southampton, July 1992. Overseas Development Administration, London, pp13–26

Allan, JA (1996) 'Policy responses to the closure of water resources: regional and global issues', in Howsam, P and Carter, R (eds), *Water Policy: Allocation and Management in Practice*, Proceedings of the International Conference on Water Policy, Cranfield University, 23–24 September 1996, E & FN SPON, London, pp3–12

Amezaga, JM and O'Connell, PE (1998) 'Unfolding the sociotechnical dimension of hydroinformatics: the role of problem structuring methods', *Hydroinformatics '98*, Balkema, Rotterdam, pp1193–1200

Arnell, NW (1996) *Global Warming, River Flows and Water Resources*, John Wiley & Sons Ltd, Chichester

Arthur, RAJ (1997) 'Water without limits', *Water and Environment*, pp16–19

Aylward, B, Echeverria, J, Fernandez Gonzalez, A F, Porras, I, Allen, K, and Mejias, R (1998) *Economic Incentives for Watershed Protection: A Case Study of Lake Arenal, Costa Rica*. Final report on a research project under the Program of Collaborative Research of Environment and Development, IIED, London

Baconguis, S (1980) 'Water balance, water use and maximum water storage of a dipterocarp forest watershed' in San Lorenzo, Norzagaray, Bulacan, Sylvatrop, *Philippines Forest Research Journal* 2, pp73–98

Bands, DP, Bosch, JM, Lamb, AJ, Richardson, DM, Van Wilgen, BW, Van Wyk, DB and Versfeld, DB (1987) *Jonkershoek Forestry Research Centre Pamphlet 384*, Department of Environment Affairs, Pretoria, South Africa

Barr, JJF (1998) 'Use of indigenous knowledge by natural resources scientists: issues in theory and practice'. Paper presented at the National Workshop on 'The State of Indigenous Knowledge in Bangladesh', Dhaka, 6–7 May, 1998

Barr, JJF and Gowing, JW (1998) 'Rice production in floodplains: issues for water management in Bangladesh', *Irrigation and Environment*, E & FN Spon, London

Bate, RN and Dubourg, WR (1994) *A netback analysis of water irrigation demand in East Anglia*, CSERGE Discussion Paper WM94, University College, London

Bell, JP (1976) *Neutron probe practice*, Institute of Hydrology, Report no. 19, Institute of Hydrology, Wallingford, UK

Benson, JF and Willis, KG (1991) 'The demand for forests for recreation', *Forestry Expansion: A Study of Technical, Economic and Ecological Factors*, Forestry Commission, Edinburgh, Scotland

Bhatia, R and Falkenmark, M (1992) *Water resource policies and urban poor: innovative thinking and policy imperatives*; paper presented to the Dublin International Conference on Water and the Environment, January 1992, S. Ireland

Blythe, EM, Dolman, AJ and Noilhan, J (1994) 'The effect of forest on mesoscale rainfall: An example from HAPEX-MOBILHY', *J Appl Met* 33, pp445–454

Bonell, M (1999) 'Tropical forest hydrology and the role of the UNESCO International Hydrology Programme: some personal observations'; submitted to *Hydrology and Earth System Sciences*, European Geophysical Society

Bonell, M and Gilmour, DA (1978) 'The development of overland flow in a tropical rainforest catchment', *J Hydrol*, 39, pp365–382

Bosch, JM (1979) 'Treatment effects on annual and dry period streamflow at Cathedral Peak', *S Afr For J*, 108, pp29–38

Bosch, JM and Hewlett, JD (1982) 'A review of catchment experiments to determine the effects of vegetation changes on water yield and evapotranspiration', *J Hydrol* 55, pp3–23

Bouten, W, Smart, PJF and De Water, E (1991) 'Microwave transmission, a new tool in forest hydrological research', *J Hydrol*, 124, pp119–130

Boyd, C (1997) *DFID White Paper – An ODI Perspective*, ODI website, http://www.oneworld.org/odi/

Brandt, J (1989) 'The size distribution of throughfall drops under vegetation canopies', *Catena*, 16, pp507–524

Brans, J, Vincke, P and Marescal, B (1986) 'How to select and how to rank projects: the PROMETHEE method', *European Journal of Operational Research*, vol 24, pp228–238

Bruijnzeel, LA (1990) *Hydrology of moist tropical forests and effects of conversion: a state of knowledge review*, UNESCO International Hydrological Programme, Paris, France

Calder, IR (1978) 'Transpiration observations from a spruce forest and comparison with predictions from an evaporation model', *J Hydrol*, 38, pp33–47

Calder, IR (1986) 'The influence of land use on water yield in upland areas of the UK', *J Hydrol*, 88, pp201–212

Calder, IR (1990) *Evaporation in the Uplands*, John Wiley & Sons Ltd, Chichester

Calder, IR (1991) 'Implications and assumptions in using the total counts and convection-dispersion equations for tracer flow measurements – with particular reference to transpiration measurements in trees', *J Hydrol*, 125, pp149–158

Calder, IR (1992a) 'Hydrologic effects of land-use change.' Chapter 13, Maidment, DR (ed) *Handbook of Hydrology*, McGraw Hill

Calder, IR (1992b) 'A model of transpiration and growth of Eucalyptus plantation in water-limited conditions', *J Hydrol*, 130, pp1–15

Calder, IR (1992c) 'The hydrological impact of land use change (with special reference to afforestation and deforestation)', *Proceedings of the Conference on Priorities for Water Resources Allocation and Management*, Southampton, July 1992, Overseas Development Administration, London, pp91–101

Calder, IR (1996a) 'Water use by forests at the plot and catchment scale', *Commonwealth Forestry Review*, 75(1), pp19–30

Calder, IR (1996b) 'Dependence of rainfall interception on drop size: 1 Development of the two-layer stochastic model', *J Hydrol*, 185, pp363–378

Calder, IR (1997) *Capacity building support to the WRMS project*: report prepared for the government of Zimbabwe and the British Department for International Development, Ewelme, Wallingford, UK

Calder, IR (1998) *Review outline of Water Resource and Land Use Issues*, SWIM Paper 3, International Irrigation Management Institute, Colombo, Sri Lanka

Calder, IR, and Bastable, HGD (1995) *Comments on the Malawi Government Water Resources Management Policy and Strategies*, report to ODA, Institute of Hydrology, Wallingford, UK

Calder, IR, Hall, RL and Prasanna, KT (1993) 'Hydrological impact of Eucalyptus plantation in India', *J Hydrol*, 150, pp635–648

Calder, IR, Hall, RL, Bastable, HG, Gunston, HM, Shela, O, Chirwa, A and Kafundu, R (1995) 'The impact of land use change on water resources in sub-Saharan Africa: a modelling study of Lake Malawi', *J Hydrol*, 170, pp123–135

Calder, IR, Hall, RL, Rosier, PTW, Bastable, HG and Prasanna, KT (1996) 'Dependence of rainfall interception on drop size: 2 Experimental determination of the wetting functions and two-layer stochastic model parameters for five tropical tree species', *J Hydrol*, 185, pp379–388

Calder, IR, Harding, RJ and Rosier, PTW (1983) 'An objective assessment of soil-moisture deficit models', *J Hydrol*, 60, pp329–355

Calder, IR and Newson, MD (1979) 'Land use and upland water resources in Britain – a strategic look', *Water Resources Bulletin*, 16, pp1628–1639

Calder, IR. and Newson, MD (1980) 'The effects of afforestation on water resources in Scotland. Land assessment in Scotland'. Proceedings of the Royal Scottish Geographical Society, Edinburgh, May 1979, Aberdeen University Press, Aberdeen, pp 51–62

Calder, IR, Newson, MD and Walsh, PD (1982) 'The application of catchment, lysimeter and hydrometeorological studies of coniferous afforestation in Britain to land-use planning and water management', *Proc Symp Hydrolog Research Basins*, Bern 1982, pp853–863

Calder, IR and Rosier, PTW (1976) 'The design of large plastic sheet net-rainfall gauges', *J Hydrol*, 30, pp403–405

Calder, IR, Rosier, PTW, Prasanna, KT and Parameswarappa, S (1997a) '*Eucalyptus* water use greater than rainfall input – a possible explanation from southern India', *Hydrology and Earth System Sciences*, 1(2) pp249–256

Calder, IR, Reid, I, Nisbet, T and Robinson, MR, (1997b) *Trees and Drought Project on Lowland England*. Project proposal to the Department of the Environment, Institute of Hydrology & Loughborough University, UK

Calder, IR, Swaminath, MH, Kariyappa, GS, Srinivasalu, NV, Srinivasa Murthy, KV and Mumtaz, J (1992) 'Deuterium tracing for the estimation of transpiration from trees: 3 Measurements of transpiration from *Eucalyptus* plantation, India', *J Hydrol*, 130, pp37–48

Calder, IR and Wright, IR (1986) 'Gamma-ray attenuation studies of interception from Sitka spruce: some evidence for an additional transport mechanism', *Water Resour Res*, 22, pp409–417

Calder, IR, Wright, IR and Murdiyarso, D (1986) 'A study of evaporation from tropical rainforest – West Java', *J Hydrol*, 89, pp13–33

Cannell, MGR, Mobbs, DC and Lawson, GJ (1998) 'Complementarity of light and water use in tropical agroforests; II, Modelled theoretical tree production and potential crop yield in arid to humid climates', *For Ecol Manage*, 102, pp275–282

Carney, D (1998) 'Implementing the Sustainable Rural Livelihoods Approach', *Sustainable Rural Livelihoods, What contribution can we make?* Papers presented at the Department for International Development's Natural Resources Advisers' Conference, DFID, London

Carson, CS (1994) 'Integrated Economic and Environmental Satellite'. *Survey of Current Business*, April, pp33–49

Chapman, G, (1948) 'Size of raindrops and their striking force at the soil surface in a red pine plantation', *Eos Trans AGU*, 29, pp664–670

Chatterton, B and Chatterton, L (1996) 'Closing a water resource: some policy considerations', *Water Policy: Allocation and Management in Practice, Proceedings of the International Conference on Water Policy*, Cranfield University, 23–24 September, E & FN Spon, pp355–361

Chomitz, M and Kumari, K (1998) 'The domestic benefits of tropical forests: a critical review', *The World Bank Research Observer*, vol 13, 1, pp13–35

Clayton, MH and Radcliffe, J (1996) *Sustainability, A Systems Approach*, Earthscan Publications, London

Coffey, A, Holbrook, B and Atkinson, P (1996) 'Qualitative data analysis: Technologies and representations', *Sociological Research Outline*, 1(1) http://www.socresonline.org.uk/socresonline/1/1/4.html

Cohen, Y, Fuchs, M and Green, GC (1981) 'Improvement of the heat pulse method for determining sap flow in trees', *Plant, Cell and Environment*, 4, pp391–397

Collinet, J, Monteny, B and Pouyaud, B (1984) 'Le milieu physique', *Recherche et Aménagement en Mileu Forestier Tropical H: le Projet Tai de Côte d'Ivoire. Notes Techniques du MAB no 15*, UNESCO, Paris, pp35–58

Collins, RO (1990) *The Waters of the Nile, Hydropolitics and the Jonglei Canal 1900–1988*, Clarendon Press, Oxford

Cooper, JD (1980) *Measurement of water fluxes in unsaturated soil in Thetford Forest*, Institute of Hydrology, report no 66, Wallingford, UK

Cooper, JD and Kinniburgh, DG (1993) *Water resource implications of the proposed Greenwood Community Forest*, project report, National Rivers Authority, UK

Courtney, FM (1978) Personal communication

Crabtree, JR (1997) 'The supply of public access to the countryside – a value for money and institutional analysis of incentive policies', *Environment and Planning*, 29, pp1465–1476

DFID (1997) *White Paper on International Development*, Department for International Development, London

DID (1977) *Sungai Lui Representative Basin Report no 1 for 1971/72 to 1973/74; Water Resources Publication no 7*, Drainage and Irrigation Department, Kuala Lumpur, Malaysia (cited by Bruijnzeel, 1990)

DID (1986) *Sungai Tekam Experimental Basin. Transition Report July 1980 to June 1983. Water Resources Publication no 16*, Drainage and Irrigation Department, Ministry of Agriculture, Kuala Lumpur, Malaysia (cited by Bruijnzeel, 1990)

Dietrich, WE, Windsor, DM and Dunne, T (1982) 'Geology, climate, and hydrology of Barro Colorado Island', *The ecology of a tropical forest: Seasonal rhythms and long-term changes*, Smithsonian Institution, Washington DC, pp21–46

Dixon, JA, Scura, LF, Carpenter, R and Sherman, PB (1994) *Economic analysis of environmental impacts*, Earthscan Publications Ltd, London

DOE (1995) *Rural England: a nation committed to a living countryside*, Department of the Environment, The Stationary Office, London

DOE (1997) *Government response to the conclusions and recommendations of the Environment Committee: 1st report on water conservation and supply*, Department of the Environment, The Stationary Office, London

DWAF (1996) *The Working for Water Programme*, Ministry of Water Affairs and Forestry, Cape Town, South Africa

Dye, PJ (1996) 'Climate, forest and streamflow relationships in South African afforested catchments', *Commonwealth Forestry Review*, 75 (1), pp31–38

Dye, PJ and Poulter, AG (1995) 'A field demonstration of the effect on streamflow of clearing invasive pine and wattle trees from a riparian zone', *South African Forestry Journal*, 173, pp27–30

Dyer, AJ (1961) 'Measurements of evaporation and heat transfer in the lower atmosphere by an automatic eddy correlation technique', *Q J R Met Soc*, 87, pp401–412

Elkington, J (1997) *Cannibals With Forks*, Capstone Publishing Ltd, Oxford, UK

END (1998) *Water abstraction decision deals savage blow to cost-benefit analysis, Report 278*, Environmental Data Services Ltd, London[11]

Enters, T (1998) 'Methods for the economic assessment of the on-and off-site impacts of soil erosion', *Issues in Sustainable Land Management no 2*, International Board for Soil Research and Management, Bangkok

EPA (United States Environment Protection Agency) (1992) *Nonpoint Source News – Notes*, January–February 1992, Issue 18, Tesrene Institute, Alexandra, USA http://www.epa.gov/owow/info/NewsNotes/issue18/nps18con.html

ERM and University of Newcastle upon Tyne (1997) *Economic Appraisal of the Environmental Costs and Benefits of Potential Solutions to Alleviate Low Flows in*

Rivers, report to the Environment Agency, Exeter, UK

Evans, RS and Nolan, J (1989) 'A groundwater management strategy for salinity mitigation in Victorian riverine plain, Australia', *Groundwater management: Quantity and Quality*, Proceedings of the Benidorm symposium, October 1989 (IAHS publication no 188) pp487–499

Ewen, J, Parkin, G and O'Connell, PE (1999) 'SHETRAN: A coupled surface/subsurface modelling system for 3D water flow and sediment and solute transport in river basins', *ASCE Journal of Hydrologic Engineering* (in press)

Fahey, BD and Jackson, R (1997) 'Hydrological impacts of converting native forests and grasslands to pine plantations, South Island, New Zealand', *Agric For Met*, 84, pp69–82

Fahey, BD and Rowe, LK (1992) 'Land-Use Impacts', *Waters of New Zealand*, New Zealand Hydrological Society, pp265–284

Fahey, BD and Watson, AJ (1991) Hydrological impacts of converting tussock grasslands to pine plantation, Otago, New Zealand, *J Hydrol (NZ)*, 30, pp1–15

Fairhead, J and Leach, M (1996) *Misreading the African Landscape*, Cambridge University Press

Falkenmark, M (1989) 'The massive water scarcity now threatening Africa: why isn't it being addressed?' *Ambio*, 25(3), p216

Focan, A and Fripiat, JJ (1953) 'Une année d'observation de l'humidité du sol à Yangambi', *Bulletin des Séances de l'Institut Royal Colonial Belge* 24, pp971–984

Foundation for Water Research (1996) *Assessing the benefits of surface water quality improvements*, FR/CL 0005, Marlow

Ford, ED and Deans, JD (1978) 'The effects of canopy structure, stemflow, throughfall and interception loss in a young Sitka spruce plantation', *J Appl Ecol*, 15, 905–917.

Foster SSD and Grey, DRC (1997) 'Groundwater resources: balancing perspectives on key issues affecting supply and demand', *J IWEM*, 11 June, p193

Gardiner, J (1992) Integrated catchment planning and source control: A view from the NRA (paper presented at the Conflo–92 Conference, Oxford University)

Gash, JHC and Stewart, JB (1977) The evaporation from Thetford Forest during 1975, *J Hydrol*, 35, pp385–396.

Gash, JHC, Wright, IR and Lloyd, CR (1980) 'Comparative estimates of interception loss from coniferous forests in Great Britain', *J Hydrol*, 48, pp89–105

Gibbons, DC (1986) The economic value of water, *Resources for the Future*, Washington DC, USA

Gilmour, DA (1977) Logging and the environment, with particular reference to soil and stream protection in tropical rainforest situations, *FAO Conservation Guide No 1*, FAO, Rome, pp223–235

Greenwood, EAN (1992) 'Deforestation, Revegetation, Water Balance and Climate: An Optimistic Path Through the Plausible, Impracticable and the Controversial', *Adv. in Bioclimatol* Springer-Verlag, New York

Greenwood, EAN, Klein, L, Beresford, JD, and Watson, GD (1985) 'Differences in annual evaporation between grazed pasture and *Eucalyptus* species in plantations on a saline farm catchment', *J Hydrol*, 78, pp261–278

Hall, RL and Calder, IR (1993) 'Drop size modification by forest canopies – measurements using a disdrometer', *J Geophys Res*, 90, pp465–470

Hall, RL, Calder, IR, Gunawardena, ERN and Rosier, PTW (1996a) 'Dependence of rainfall interception on drop size: 3 Implementation and comparative performance of the stochastic model using data from a tropical site in Sri Lanka', *J Hydrol*, 185, pp389–407

Hall, RL, Allen, SJ, Rosier, PTW, Smith, DM, Hodnett, MG, Roberts, JM, Hopkins, R, Davies, HN, Kinniburgh, DG and Goody, DC (1996b) *Hydrological effects of short rotation coppice*, Institute of Hydrology report to the Energy Technology Support Unit, IH Wallingford, UK

Halvorson, AD and Reule, CA (1980) 'Alfalfa for hydrologic control of saline seeps', *Soil Science Society of America Journal*, 44, pp370–374

Hamilton, LS (1987) 'What are the impacts of deforestation in the Himalayas on the Ganges-Brahmaputra lowlands and delta? Relations between assumptions and facts', *Mountain Research and Development*, 7, pp256–263

Hardin, G, (1968) 'The tragedy of the commons', *Science* 162, pp1243–8

Harding, RJ, Hall, RL, Neal, C, Roberts, JM, Rosier, PTW and Kinniburgh, DK (1992) *Hydrological impacts of broadleaf woodlands: Implications for water use and water quality*, Institute of Hydrology, British Geological Survey Project Report 115/03/ST and 115/04/ST for the National Rivers Authority, Institute of Hydrology, Wallingford, UK

Harvey, DR , and Willis, K (1997) The Social Economic Value of Land, Research report to MAFF University of Newcastle upon Tyne, Newcastle upon Tyne

Hassan, R, Berns, J, Chapman, A, Smith, R, Scott, D, and Ntsaba, M (1995) *Economic policies and the environment in South Africa: the case of water resources in Mpumalanga*, Division of Forest Science & Technology, CSIR, Pretoria, South Africa

Hewlett, JD and Bosch, JM (1984) 'The dependence of storm flows on rainfall intensity and vegetal cover in South Africa', *J Hydrol*, 75, pp365–381

Hewlett, JD, and Hibbert, AR (1967) 'Factors affecting the response of small watersheds to precipitation in humid areas', *International Symposium on Forest Hydrology*, Pergamon Press, Oxford, pp275–90

Hewlett, JD and Helvey, JD (1970) 'Effects of forest clearfelling on the storm hydrograph', *Water Resour Res*, 6 (3), pp768–782

Hingston, FJ and Gailitis, V (1976) 'The geographic variation of salt precipitated over Western Australia', *Aust J Soil Res*, 14, pp319–335

House of Commons. Environment Committee (1996) 1st report, session 1996–97: water conservation and supply, The Stationary Office, London

Howell, P, Lock, M and Cobbs, S (1988) *The Jonglei Canal: Impact and Opportunity*, Cambridge University Press, Cambridge

Hudson, JA and Gilman, K (1993) 'Long-term variability in the water balances of the Plynlimon catchments', *J Hydrol*, 143, pp355–380

Hummel, FC (1992) 'Aspects of forest recreation in Western Europe', *Forestry*, 65(3), pp237–251

Huttel, C (1975) Recherches sur l'écosystème de la forêt subéquatoriale de basse Côte d'Ivoire: IV Estimation du bilan hydrique, *La Terre et la Vie* 29, pp192–202

ICRAF (1994) *Annual Report*, International Centre for Research in Agroforestry, Nairobi, Kenya

IUCN (1998) *The Green Accounting Initiative* http://iucn.org/places/usa/gai-activitiespage.html

IUCN, UNEP and WWF (1991) *Caring for the Earth, A Strategy for Sustainable Living*, Earthscan, London

Institute of Hydrology (1994) *Amazonia: Forest, pasture and climate – results from ABRACOS*, Institute of Hydrology, Wallingford, UK

Jamieson, DG (ed) (1996) 'Special Issue: Decision-Support Systems', *J Hydrol*, 177

Jamieson, DG and Fedra, K (1996) 'The 'WaterWare' decision-support system for river-basin planning: 1 Conceptual design', *J Hydrol*, 177, pp163–175

Johnson, RC (1995) *Effects of upland afforestation on water resources: the Balquhidder experiment 1981–1991*, report no 116, Institute of Hydrology, Wallingford, UK

Jones, JA and Grant, GE (1996) 'Peak flow responses to clear-cutting and roads in small and large basins, western Cascades, Oregon', *Water Resour Res*, 32, pp959–974

Kilsby, CG, Ewen, J, Sloan, WT and O'Connell, PE (1998) 'Modelling the hydrological impacts of climate change at a range of scales', *Proceedings of the Second International Conference on Climate and Water*, Espoo, Finland, vol 3, pp1402–1411

Kirby, C, Newson, MD and Gilman, K (1991) *Plynlimon research: the first two decades*, report no 109, Institute of Hydrology, Wallingford, UK

Kline, JR, Martin, JR, Jordan, CF and Koranda, JJ (1970) 'Measurement of transpiration in tropical trees with tritiated water', *Ecology*, 51, pp1068–1073

Kutcher, G, McGurk, S and Gunaratnam, J (1992) *China: Yellow River Basin*. Water investment planning study presented at a World Bank irrigation and drainage seminar (cited by Winpenny, 1996)

Langford, KJ (1976) Change in yield of water following a bushfire in a forest of *Eucalyptus regnans*, *J Hydrol*, 29, pp87–114

Law, F (1956) 'The effect of afforestation upon the water yield of water catchment areas', *J Br Waterworks Assoc*, 38, pp489–494.

Leach, M and Mearns, R (1996) Environmental change and policy – Challenging received wisdom in Africa, *The Lie of the Land*, Villiers Publication, London, pp1–33

Ledger, DC (1975) The water balance of an exceptionally wet catchment area in West Africa, *J Hydrol*, 24, pp207–214

Le Maitre, DC, Van Wilgen, BW, Chapman, RA and McKelly, DH (1996) 'Invasive plants and water resources in the Western Cape Province, South Africa: modelling the consequences of a lack of management', *J Appl Ecol*, 33, pp161–172

Leopoldo, PR, Franken, W, Matsui, E and Salati, E (1982a) Estimation of evapotranspiration of 'terra firme' Amazonian forest, *Acta Amazonica*, 12, pp23–28 (in Portuguese with English summary)

Leopoldo, PR, Franken, W and Salati, E (1982b) Water balance of a small catchment area in 'terra firme' Amazonian forest, *Acta Amazonica*, 12, pp333–337 (in Portuguese with English summary)

Leyton, L, Reynolds, ERC and Thompson, FB (1967) Rainfall interception in forest and moorland. International Symposium on Forest Hydrology, Pergamon Press, Oxford, pp163–168

Lill, WS van, Kruger, FJ and Van Wyk, DB (1980) The effects of afforestation with *Eucalyptus grandis* Hill ex Maiden and *Pinus patula* Schlecht et Cham on streamflow from experimental catchments at Mokobulaan, Transvaal, *J Hydrol*, 48, pp107–118

Loucks DP and Gladwell, JS (1999) *Sustainability Criteria for Water Resource Systems*, Cambridge University Press, Cambridge

Low, KS and Goh, GC, (1972) 'The water balance of five catchments in Selangor, West Malaysia', *Journal of Tropical Geography*, 35, pp60–66

Luvall, JR and Murphy, CE (1982) 'Evaluation of the tritiated water method for measurement of transpiration in young *Pinus taeda*', *Forest Sci*, 28, pp5–16

Macumber, P (1990) The salinity problem, *The Murray*, Murray-Darling Basin Commission, Canberra, Australia, pp111–125

Malcolm, CV (1990) Saltland agronomy in Western Australia – an overview, *Revegetation of Saline Land*, Proceedings of a workshop held at the Institute for Irrigation and Salinity Research, Tatura, Victoria, Australia

Marshall, JS and Palmer, WM (1948) 'The distribution of raindrops with size', *J Meteorology*, 5, pp165–166

Meadows, DH, Meadows, DL, Randers, J, and Behrens, WW (1972) *The Limits to Growth; A Report for the Club of Rome's Project on the Predicament of Mankind*, Universe Books, New York, USA

Meher-Homji, VM (1980) Repercussions of deforestation on precipitation in Western Karnataka, India, *Aech Met Geogph Biokl*, Series B 28, pp385–400

Milburn, A (1997) The need for a blue revolution in the fresh water sector, *Proceedings of the First World Water Forum, Marrakesh, Morocco*, Elsevier, pp37–43

Miller, BJ (1994) *Soil water regimes of the Glendhu Experimental Catchments*; unpublished dissertation, Department of Geography, University of Otago, New Zealand

Mitchell, JK and Bubenzer, GD (1980) Soil loss estimation, in Kirkby MJ and Morgan RPC (eds), *Soil Erosion* Wiley, Chichester, USA, pp17–62

Molden, D (1997) *Accounting for water use and productivity*, SWIM Paper 1 International Irrigation Management Institute, Colombo, Sri Lanka

Monteith, JL (1965) 'Evaporation and environment', *Symp Soc Exper Biol*, 19, pp205–234

Morgan, RPC, Morgan, DDV and Finney, HJ (1984) A predictive model for the assessment of soil erosion risk, *Journal of Agricultural Engineering Research* 30, pp245–53

Mosely, MP (1988) *Climate Change Impacts – The Water Industry*, CH 20, Ministry for the Environment, Wellington, New Zealand

Nacario-Castro, E (1997) *When the well runs dry: A civil initiative in watershed planning and management in the Philippines*, Ramon Aboritiz Foundation Inc, Cebu, Philippines

Newson, MD (1990) 'Forestry and water 'good practice' and UK catchment policy', *Land Use Policy*, 7:1, pp53–58

Newson, MD (1991) 'Catchment control and planning: emerging patterns of definition, policy and legislation in UK water management', *Land Use Policy*, 9:1, pp9–15

Newson, MD (1992a) *Land, water and development*, Routledge, London

Newson, MD (1992b) Land and water: convergence, divergence and progress in UK policy, *Land Use Policy*, 9:2, pp111–121

Newson, MD (1997) *Land, water and development*, Routledge, London

Newson, MD, Gardiner, J and Slater, S (1998) 'Planning and managing for the future', *The Changing Hydrology of the UK*, Routledge (in press), London

Noordwijk, M van, Lawson, G, Soumaré, A, Groot, JJR and Hairiah, K (1996) Root distribution of trees and crops: competition and/or complementarity; in Ong, CK and Huxley, P (eds) *Tree–Crop Interactions*, CAB International, Wallingford, UK, pp319–364

O'Callaghan, JR (1995) 'NELUP: An introduction', *J Environ Plann Manage*, 38(1)

O'Callaghan, JR (1996) *Land Use. The interaction of economics, ecology and hydrology*, Chapman & Hall, London

O'Connell, PE (1995) 'Capabilities and limitations of regional hydrological models', *Scenario Studies for the Rural Environment*, Kluwer Academic Publishers, The Netherlands, pp143–156

Oldeman, LR, Hakkeling, RTA and Sombroek, WG (1991) Second revised edition. *World Map of the Status of Human-Induced Soil Degradation: An explanatory note*, International Soil Reference and Information Centre, Wageningen, The Netherlands

Olszyczka, B (1979) *Gamma-ray determinations of surface water storage and stem water content for coniferous forests*, PhD thesis, Department of Applied Physics, University of Strathclyde

Omerod, RJ (1996) Informations systems strategy development at Sainsbury's supermarkets using 'soft' OR, *Interfaces*, 26:1, pp102–130

Ong, CK, Odango, JCW, Marshall, F and Black, CR (1991) 'Water use by trees and crops. Five hypotheses', *Agroforestry Today*, April–June, pp7–10

Ong, CK, Black, CR Marshall, F and Corlett, JE (1996) Principles of resource capture and utilization of light and water; in Ong, CK and Huxley, P (eds) *Tree-Crop Interactions*, CAB International, Wallingford, UK, pp73–158

Ong, CK (1996) A framework for quantifying the various effects of tree-crop interactions; in Ong, CK and Huxley, P (eds) *Tree-Crop Interactions*, CAB International, Wallingford, UK, pp1–23

Pallett, J (ed) (1997) *Sharing water in southern Africa*, Desert Research Foundation of Namibia, Windhoek, Namibia

Panayotou, T and Hupe, K (1996) *Environmental Impacts of Structural Adjustment Programmes: Synthesis and Recommendations*, UNEP, Environmental Economics Series, paper no 21, Nairobi, Kenya

Pearce, AJ (1986) *Erosion and sedimentation* (working paper) Environment and Policy Institute, Honolulu, Hawaii

Penman, HL (1948) 'Natural evaporation from open water, bare soil and grass', *Proc Roy Soc Ser A*, 193, pp120–145

Penman, HL (1963) *Vegetation and hydrology*, tech comm 53, Commonwealth Bureau of Soils, Harpenden, UK

Pereira , HC (1989) *Policy and practice in the management of tropical watersheds*, Westview Press, Colorado, USA

Pereira, HC, McCulloch, JSG, Dagg, M, Kerfoot, O, Hosegood, PH and Pratt, MAC (1962) 'Hydrological effects of changes in land use in some E. African catchment areas', *East African Agricultural and Forestry Journal* 27 (Special Issue)

Poels, R (1987) *Soils, water and nutrients in a forest ecosystem in Surinam*, PhD thesis, Agricultural University, Wageningen, The Netherlands (cited by Bruijnzeel, 1990)

Price, DJ, Calder, IR and Johnson, RC (1995) *Modelling the effect of upland afforestation on water resources*, report to the Scottish Office, Institute of Hydrology, Wallingford, UK

Price, DJ, Calder, IR, Shela, O, Chirwa, A and Williams, H (1998) Investigation into the impact of changing forest cover upon the water resources of the Zambezi Basin; submitted to *Tree Physiology*

Ranganathan, R and de Wit, CT (1996) Mixed cropping of annuals and woody perennials: an analytical approach to productivity and management; in: Ong, CK and Huxley, P (eds) *Tree-Crop Interactions*, CAB International, Wallingford, UK, pp25–49

Rao, MR, Sharma, M and Ong, CK (1990) A study of the potential of hedgerow intercropping in semiarid India using a two-way systematic design, *Agrofor Syst* 11, pp243–258

Rees, JA, Williams, S, Atkins, JP, Hammond, CJ and Trotter, SD (1993) *Economics of water resource management*, R&D note 128, National Rivers Authority, Bristol, UK

Reid, I and Parkinson, RJ (1984) The nature of the tile-drain outfall hydrograph in heavy clay soils, *J Hydrol* 72, pp289–305

Roberts, JM (1977) The use of 'tree cutting' techniques in the study of the water relations of mature *Pinus sylvestris* (L): The technique and survey of the results, *J Exper Bot*, 28, pp751–767

Roberts, JM (1978) 'The use of the 'tree cutting' technique in the study of the water relations of Norway spruce' (*Picea abies* (L) Karst), *J Exper Bot*, 29, pp465–471

Robertson, G (1996) Saline Lands in Australia: extent and predicted trends, *Proceedings of the 4th National Conference and Workshop on the Productive Use and Rehabilitation of Saline Lands*, Promaco Conventions PTY Ltd, Australia

Robinson, M, Moore, RE and Blackie, JR (1997) *From Moorland to Forest: The Coalburn Catchment Experiment*, Institute of Hydrology and Environment Agency Report, Institute of Hydrology, Wallingford, UK

Robinson, M, Ryder, EL, and Ward, RC (1985) Influence on streamflow of field drainage in a small agricultural catchment, *Agricultural Water Management*, 10, pp145–158

Roche, MA (1982) Evapotranspiration réelle de la forêt amazonienne en Guyane, Cahiers ORSTOM, série Hydrologie, 19, pp37–44

Rosenhead, J (1996) What's the problem? An introduction to problem structuring methods, *Interfaces*, 26 (6), pp117–131

Rowntree, PR (1988) Review of general circulation models as a basis for predicting the effects of vegetation change on climate; in Reynolds, ERC and Thompson, FB (eds) *Forests, Climate and Hydrology: Regional impacts*, Kefford Press, UK, pp162–193

Rosier, PTW (1987) Personal communication

Rutter, AJ (1963) 'Studies in the water relations of *Pinus sylvestris* in plantation conditions': Measurements of rainfall and interception, *J Ecol*, 51, pp191–203

Rutter, AJ, Kershaw, KA, Robins, PC and Morton, AJ (1971) A predictive model of rainfall interception in forests: Derivation of the model from observations in a plantation of Corsican pine, *Agric Meteorology*, 9, pp367–384

Sanderson, RA (1998) VIPER – Vegetation Investigation Program for ESA Research. Users Guide and Reference Manual, Ministry of Ag. Fisheries and Food, Nobel House, London

Sanderson, RA, and Rushton, SP (1995) 'VEMM: Predicting the effects of agricultural management and environmental conditions on semi-natural vegetation', *Computers and Electronics in Agriculture*, 12, pp237–247

Scoones, I (1998) *Sustainable Rural Livelihoods: A Framework for Analysis* (working paper no 72), Institute of Development Studies, Brighton, UK

Scott, DF (1993) 'The hydrological effects of fire in South African mountain catchments', *J Hydrol*, 150, pp409–432

Scott, DF and Lesch, W (1997) 'Streamflow responses to afforestation with *Eucalyptus grandis* and *Pinus patula* and to felling in the Mokobulaan experimental catchments, South Africa', *J Hydrol*, 199, pp360–377

Scott, DF and Smith, RE (1997) 'Preliminary empirical models to predict reduction in total and low flows resulting from afforestation', *Water SA*, 23, pp135–140

Seckler, D (1996) *The new era of water resources management*, research report 1, International Irrigation Management Institute, Colombo

Sherriff, J (1996) 'Water resources management in England and Wales'; in Howsam, P and Carter, R (eds), *Water Policy: Allocation and Management in Practice, Proceeding of the International Conference on Water Policy*, Cranfield University, E & FN Spon, London, pp68–69

Shiklomanov, IA (1999) Climate change hydrology and water resources: the work of the IPCC, 1988–1994; in Van Dam, JC (ed) *Impacts of Climate Change and Climate Variability on Hydrological Regimes*, UNESCO International Hydrology Series, Cambridge University Press, Cambridge, pp8–20

Shuttleworth, WJ (1988) Evaporation from Amazonian Rainforest, *Proc R Soc Lond B*, 233, pp321–346

Sillitoe, P, Dixon, P and Barr, J (1998) 'Indigenous knowledge research on the floodplains of Bangladesh: the search for a methodology', *Grassroots Voice*, 1(1), ISSN 1560-358X

Simmons, P, Poulter, D and Hall, NH (1991) *Management of Irrigation Water in the Murray-Darling Basin* (discussion paper 91.6), Australian Bureau of Agricultural and Resource Economics, Canberra

Simon, HA (1969) *The Sciences of the Artificial*, MIT Press, Cambridge

Simonovic, SP and Bender, MJ (1996) 'Collaborative planning-support system: an approach for determining evaluation criteria', *J Hydrol*, 177, pp237–251

Slater, S, Newson, MD and Marvin, SJ (1995) 'Land use planning and the water sector: a review of development plans and catchment management plans', *Town Planning Review*, 65(4), pp375–397

Smith, EJ (1997) 'The balance between public water supply and environmental needs', *J IWEM*, 11 February, pp8–13

Stebbing, EP (1937) The threat of the Sahara, *Journal of the Royal African Society*, Extra Supplement, May, pp3–35

Stocking, M (1996) 'Soil erosion – breaking new ground', *The Lie of the Land*, Leach, M and Mearns, R (eds) Villiers Publication, London, pp140–154

Streeter, BA (1997) Tradable rights for water abstraction, *J IWEM* 11, August, pp277–281

Sutcliffe, JV (1974) A hydrological study of the Southern Sudd region of the Upper Nile, *Hydrological Sciences Bulletin*, 19, pp237–55

Swift, J (1996) Desertification – Narratives, Winners and Losers, Leach, M and Mearns, R (eds) *The Lie of the Land*, Villiers Publication, London, pp140–154

Tañada, CR (1997) CUSW (A): *Mobilizing for Sustainable Water Resources*, The Asian Institute of Management, Eugenio López Foundation, Cebu

Taylor, A and Patrick, M (1987) 'Looking at water through different eyes – the Maori Perspective', *Soil and Water*, pp2–24

Taylor, CH and Pearce, AJ (1982) Storm runoff processes and sub-catchments characteristics in a New Zealand hill country catchment, *Earth Surf Process Land forms*, 7, pp439–447

Tiffen, M, Mortimore, M and Gichuki, F (1993) *More People, Less Erosion*, John Wiley & Sons Ltd, Chichester

Tolentino, AS (1996) Legal and institutional aspects of groundwater development in the Philippines; in Howsam, P and Carter, R (eds) *Water Policy: Allocation and Management in Practice*, Proceedings of the International Conference on Water Policy, Cranfield University, E & FN Spon, London, pp283–289

Tuinder, BA den, Calder, IR, Helland-Hansen, E, Koziorowski, G, Murungweni, Z, Chidenga, E, Mujuru, L, Mutamiri, J, and Bowen-Williams, H (1995) *Programme for the development of a National Water Resources Management Strategy (WRMS) for Zimbabwe*, Ministry of Lands and Water Resources, Government of Zimbabwe

United Nations Department for Policy Coordination and Sustainable Development (1997) *Comprehensive Assessment of the Freshwater Resources of the World*, Commission on Sustainable Development, New York

US Department of Agricultural Research Service (1961) *A Universal equation for Predicting Rainfall-erosion Losses*, USDA-ARS special report, pp22–26

Van Lill, WS, Kruger, FJ and Van Wyk, DB (1980) 'The effects of afforestation with Eucalyptus grandis Hill ex Maiden and Pinus patula Schlecht. Et Cham. on stream-flow from experimental catchments at Mokobulaan, Transvaal', *J Hydrol*, 48, pp107–118

van Noordwijk, M, Lawson, G, Soumaré, A, Groot, JJR and Hairiah, K (1996) 'Root distribution of trees and crops: competition and/or complementarity', in Ong, CK and Huxley, P (eds) *Tree-crop Interactions*, CAB International, Wallingford, pp319–364

Versfeld, DB (1981) 'Overland flow on small plots at the Jonkershoek Forestry Research Station', *South African Forestry Journal*, 119

Versfeld, DB and Wilgen, BW van (1986) Impacts of woody aliens on ecosystem properties; in Macdonald, IAW, Kruger, FJ and Ferrar, AA (eds) *The Ecology and Control of Biological Invasions in South Africa*, Oxford University Press, Cape Town, South Africa, pp239–246

Walker, DH and Sinclair, FL (1998) 'Acquiring qualitative knowledge about complex agroecosystems; Part 2: Formal representation', *Agricultural Systems* 56(3), pp365–386

Walling, DE (1983) The sediment delivery problem, *J Hydrol*, 65, pp209–237

Waugh, J (1992) Introduction: Hydrology in New Zealand; in Mosely, MP (ed) *Waters of New Zealand*, New Zealand Hydrological Society, pp1–12

Western Australian Department of Agriculture (1988) *Salinity in Western Australia – A Situation Statement*, Technical Report no 81, Division of Resource Management, Western Australian Department of Agriculture Perth

Wicht, CL (1939) Forest influence research techniques Jonkershoek, *Journal of the South African Forestry Association* 3, pp65–80

Willis, KG and Benson, JF (1989) Recreational values of forests, *Forestry*, 62(2), pp93–110

Winpenny, JT (1992) 'Water as an economic resource', *Proceedings of the Conference on Priorities for Water Resources Allocation and Management*, Southampton, July, Overseas Development Administration, London, pp35–41

Winpenny, JT (1996) The value of water valuation; in Howsam, P and Carter, R (eds) *Water Policy: Allocation and Management in Practice*, *Proceeding of the International Conference on Water Policy*, Cranfield University, E & FN Spon, London, pp197–204

Wischmeier, WH and Smith, DD (1965) *Predicting Rainfall Erosion from Cropland East of the Rocky Mountains*, Agricultural Handbook no 282, United States Department of Agriculture, Washington DC, USA

Wood, WE (1924) Increase of salt in soil and streams following the destruction of the native vegetation, *J Royal Soc Western Australia*, 10, pp35–47

World Commission on Environment and Development (1987) *Our Common Future*, Oxford University Press, New York

Xue, Y (1997) Biosphere feedback on regional climate in tropical north Africa, *Quart J Roy Meteorol Soc*, 123, pp1483–1515

Index